Healing Springs of Russia

Svetlana Malkhazova · Dmitry Orlov ·
Natalia Shartova · Sergey Starikov ·
Tatiana Puzanova

Healing Springs of Russia

Springer

Svetlana Malkhazova
Faculty of Geography
Lomonosov Moscow State University
Moscow, Russia

Natalia Shartova
Faculty of Geography
Lomonosov Moscow State University
Moscow, Russia

Tatiana Puzanova
Faculty of Geography
Lomonosov Moscow State University
Moscow, Russia

Dmitry Orlov
Faculty of Geography
Lomonosov Moscow State University
Moscow, Russia

Sergey Starikov
Department of Medical Rehabilitation
Medical Institute of Continuing Education
MGUPP
Moscow, Russia

ISBN 978-3-030-83533-0 ISBN 978-3-030-83534-7 (eBook)
https://doi.org/10.1007/978-3-030-83534-7

Partial translation from the Russian language edition: *Медико-географический атлас России «Целебные источники и растения»* by Svetlana Malkhazova, et al., © Lomonosov Moscow State University 2019. Published by Lomonosov Moscow State University. All Rights Reserved.

This Springer imprint is published by the registered company Springer Nature Switzerland AG
The registered company address is: Gewerbestrasse 11, 6330 Cham, Switzerland

Preface

Advances in scientific medicine have not diminished the importance of healing natural remedies, the use of which is rooted in ancient times. On the contrary, modern theories of medicine and healthcare, viewed through the prism of existing economic and ecological problems, increase interest in natural healing resources and the age-old experience with them in traditional medicine.

Mineral waters have a physiological effect on the human body because of their biologically active healing properties. They are the main therapeutic agent used at balneological resorts, where they are applied externally in baths and pools, and for inhalation, drinking, and other procedures. Therapeutic mud (peloids) is another important natural healing factor widely used in Russia. The healing properties of the mud are due to its chemical composition, biologically active substances, and thermal and mechanical effects.

This book is a result of the collective efforts of members of the Lomonosov Moscow State University Faculty of Geography, with the assistance of colleagues from related organizations. It was conceived to mark the 300th anniversary of the first exploration of healing springs in Russia. This is the first cartographic synthesis at the national level of comprehensive information about natural healing resources such as healing waters and mud. The database contains information about springs, peloid deposits, and health resorts. Sources of information were cartographic, literary, and archival, along with stock materials, state statistics, and the results of field expeditions. The material sent to us by regional organizations in response to inquiries made through the Russian Geographical Society were of particular importance. The authors are sincerely grateful to all of them.

The monograph summarizes a variety of information about the natural healing resources of Russia at the federal and regional levels, presenting it in a systematic and visual form. It contains 36 maps. Along with the overview maps of Russia, the book contains maps of some regions (Caucasus, Tyva, Far East, etc.), prepared from original research in conjunction with the relevant regional services. Color illustrations and photographs complement text and maps.

The book is intended for the use of workers in the healthcare system, science, and education, as well as a wide range of users interested in natural healing resources and a healthy lifestyle.

Moscow, Russia

Prof. Svetlana Malkhazova
Head of the Department of
Biogeography, Faculty of Geography

Acknowledgments

This book was based on Медико-географический атлас России «Целебные источники и растения» (Medico-geographical Atlas of Russia "Healing Springs and Plants") by Svetlana Malkhazova, et al., published by Lomonosov Moscow State University in 2019.

The Medico-Geographical Atlas of Russia: Healing Springs and Plants was created by collecting, analyzing, and summarizing the stock materials relating to healthcare and tourism organizations with the financial support of the Russian Geographical Society, grants No. 02/2016-I of 04/05/2016, No. 02/2017-I of 06/23/2017 and No. 01/2018-I of 12/12/2018.

The authors wish to thank their colleagues, graduate students, and the students of the Faculty of Geography of Moscow State University for their professional advice, assistance in collecting and processing material, especially to L. V. Ezerova, T. V. Kotova, N. V. Ryabova for maps preparing.

Invaluable help was provided by materials received from the Russian regions:

FSBI Pyatigorsk State Research Institute of Resort Medicine of the Federal Medical and Biological Agency;
State Unitary Enterprise of Republic of Crimea "Crimean Hydrogeological Operational Station," especially lead geologist V.I. Vasenko;
Kara-kys Arakchaa and employees of the Research Institute of Medical and Social Problems and Management of Tyva Republic of the Ministry of Health of Tyva Republic, the Laboratory of Arzhaanology and Tourism LLC, and the Tyvan Republican Branch of the Russian Geographical Society;
Ministry of International Cooperation and Foreign Economic Relations of Transbaikal region
Ministry of Health of Arkhangelsk Oblast and Solonikha resort;
State Committee for Nature Management and Environmental Protection of Pskov Oblast and Khilovo resort;
Ministry of Economic Development of Tver Oblast and Kashin resort;
Lake Karachi Resort LLC.

RUSSIAN GEOGRAPHICAL SOCIETY
FOUNDED IN 1845

Contents

Introduction

Modern areas of medicine and health care contribute to increasing interest in natural healing resources and their centuries-old use in traditional medicine. The issue of using healing resources in medical practice and health care is closely related to its informational support through databases, cadastres, reference books, and cartographic models of various types, purposes, territorial coverage, and scale.

Based on the processing and systematization of the materials collected, a relational database with the spatial references of all objects under study has been prepared. After the processing of statistical information, maps were compiled in ESRI, ArcGIS, Bently's, MicroStation, and Adobe Illustrator.

The preparation process of the monograph included several stages, and shaped around the following tasks:

- a review of historical experience using medicinal springs and mud in this country and abroad;
- a systematization and analysis of data on the role of medicinal springs and mud in modern scientific and practical medicine;
- creation of a database of healing waters and mud deposits;
- mapping of the location of major healing waters and mud deposits, according to their physicochemical and therapeutic characteristics;
- development and generalization of recommendations on the use of healing waters and mud deposits in relation to certain categories of diseases.

The purpose of physical and administrative structure maps of the Russian Federation (Maps I, II) is to serve as a basis for linking thematic content of subsequent maps to the geographical objects and 85 administrative units of the Russian Federation.

Natural conditions largely determine the territorial organization and living conditions of the population, its economic activities, as well as the availability of therapeutic and recreational resources: mineral springs and peloids.

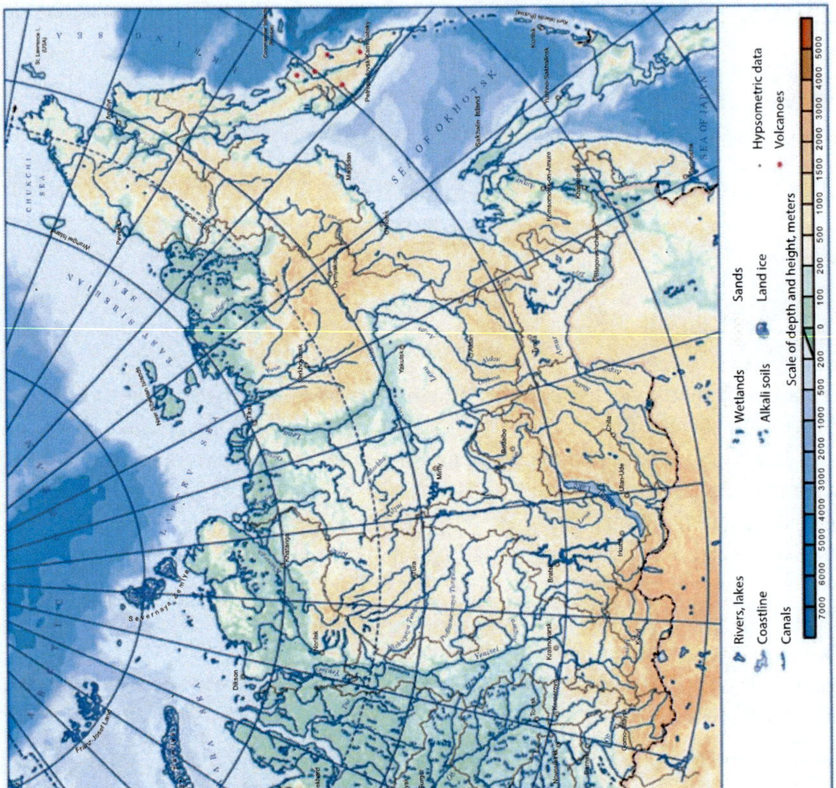

Map I Physical map

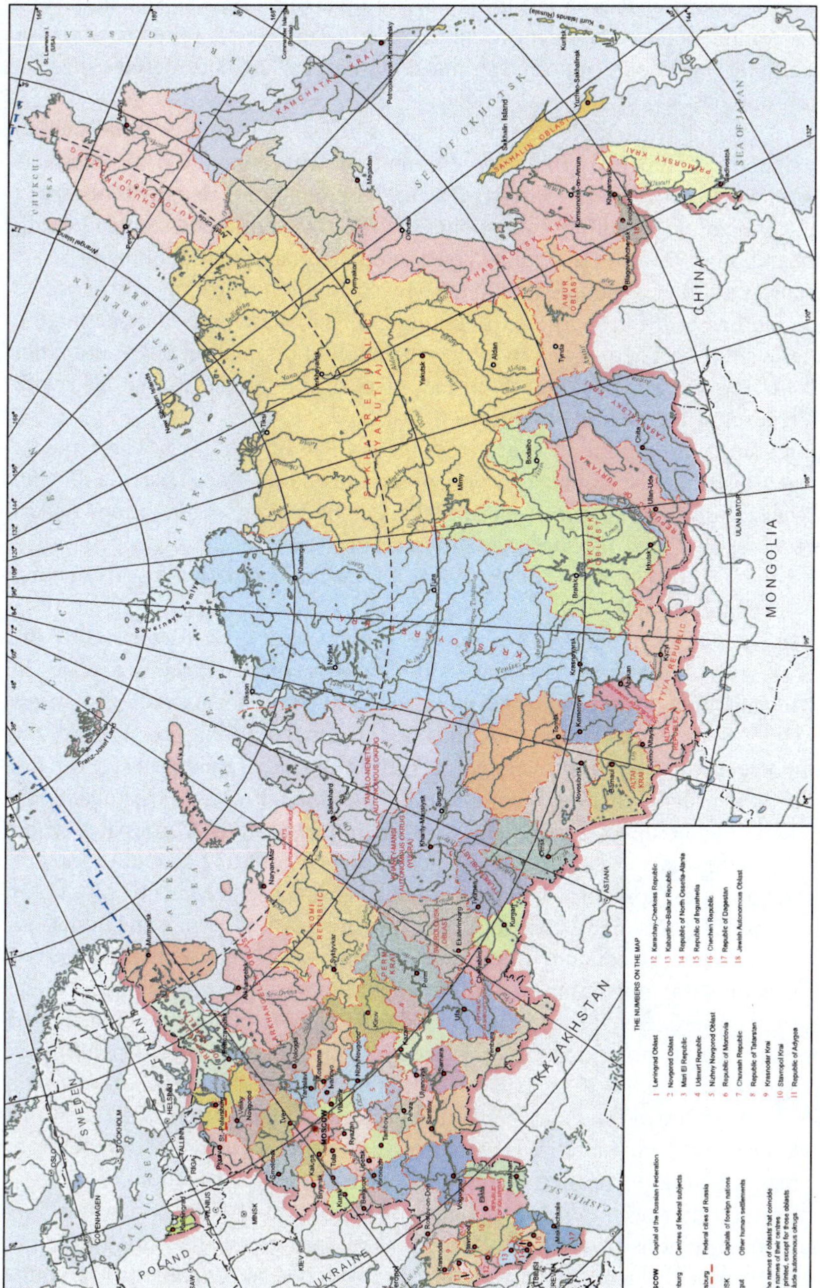

Map II Administrative divisions of the Russian Federation

The large latitudinal (more than 4,000 km) and meridional (more than 10,000 km) extent of the territory of Russia led to its natural diversity. Landscapes, as territories with a homogeneous geological foundation, a homogeneous relief, a common climate, a regular set of soils, plant and animal communities, are considered a specific human environment.

For effective and reasonable use of healing springs, it is highly important to be aware of the specifics of population distribution. The population of Russia is extremely unevenly distributed (Map III). The highest population density is observed in Central Russia, then it declines incrementally toward the north and south. Overall, the European part of the country can be considered the main population band.

The greatest number of people live in the Moscow region (Moscow and Moscow Oblast), as well as St. Petersburg and Leningrad Oblast, which are large economic centers of the country. There is an obvious association of population distribution not only with the level of socioeconomic development of the territories, but also with the comfort level of the natural conditions.

The potential healing mineral water resources of Russia are great. The types of mineral water distributed in country are very diverse in quality. The close relationship between water chemistry, composition of rocks, and hydrological conditions makes it possible to divide them into the large groups represented on the hydrogeological map "Mineral water provinces" (Ivanov 1968), where characteristic types of mineral water are presented (Map IV).

The map shows five mineral water provinces. Province I is characterized by the formation of strongly overheated mainly nitrogenous-carbonated thermal water in modern volcanic foci, which, when it reaches the surface, forms boiling springs and powerful gas-vapor and steam-water jets. In the upper oxidizing zone, under the direct influence of volcanic gases, very acidic, sulfurous-carbonated water is formed. Carbonated hot springs are quite common in the province. Province II is characterized by diverse and complex geological and structural conditions. Carbonated water prevails here. If there is nitrogen, methane, and hydrogen sulfide in them, nitrogenous-carbonated, methane-carbonated water is formed, and water with even more complex gas composition. In Province III, nitrogenous mineralized hot springs are common. Hot springs are often associated with granitoid rock massifs and with water rising from great depths and appearing as high-temperature springs. In Province IV, the distribution of nitrogenous, nitrogenous-methane, and methane water is subordinated to the general hydrochemical zonality, which can get more complicated depending on the lithological rock composition, paleo-hydrogeological, and other conditions. The most important feature of the gas composition of the province's water is the presence of hydrogen sulfide. The water of Province V is characterized by low temperature and low mineralization increasing in cases of its enrichment with radon.

The monography presents the geography of springs in Russia as a whole and for its individual regions (mainly for large areas of spring concentration or for the oldest resorts). For the Russian territory, it is developed and detailed in a series of analytical maps grouped in two main sections: Mineral water and Mud springs (peloids), indicating the locations of various types of deposits. In a separate chapter,

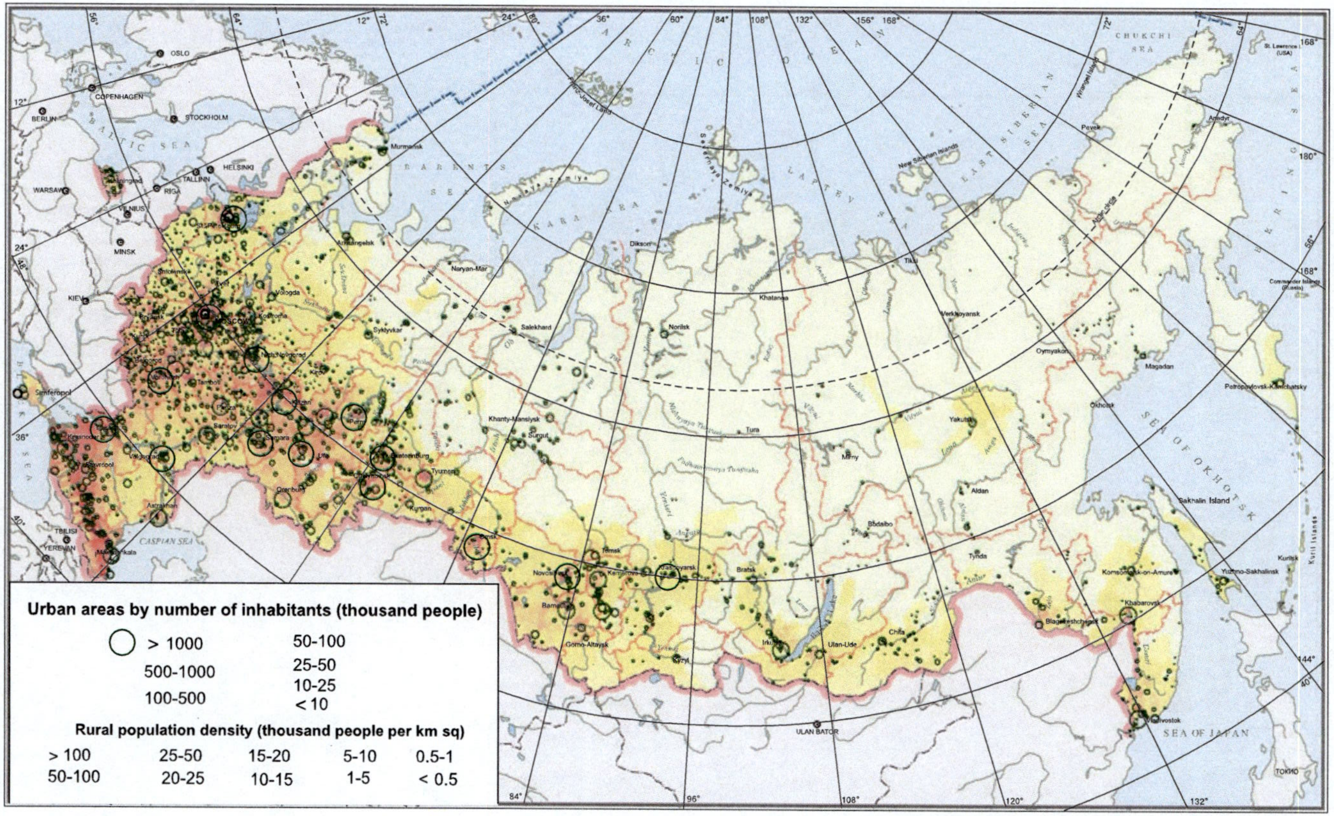

Map III Urban and rural population

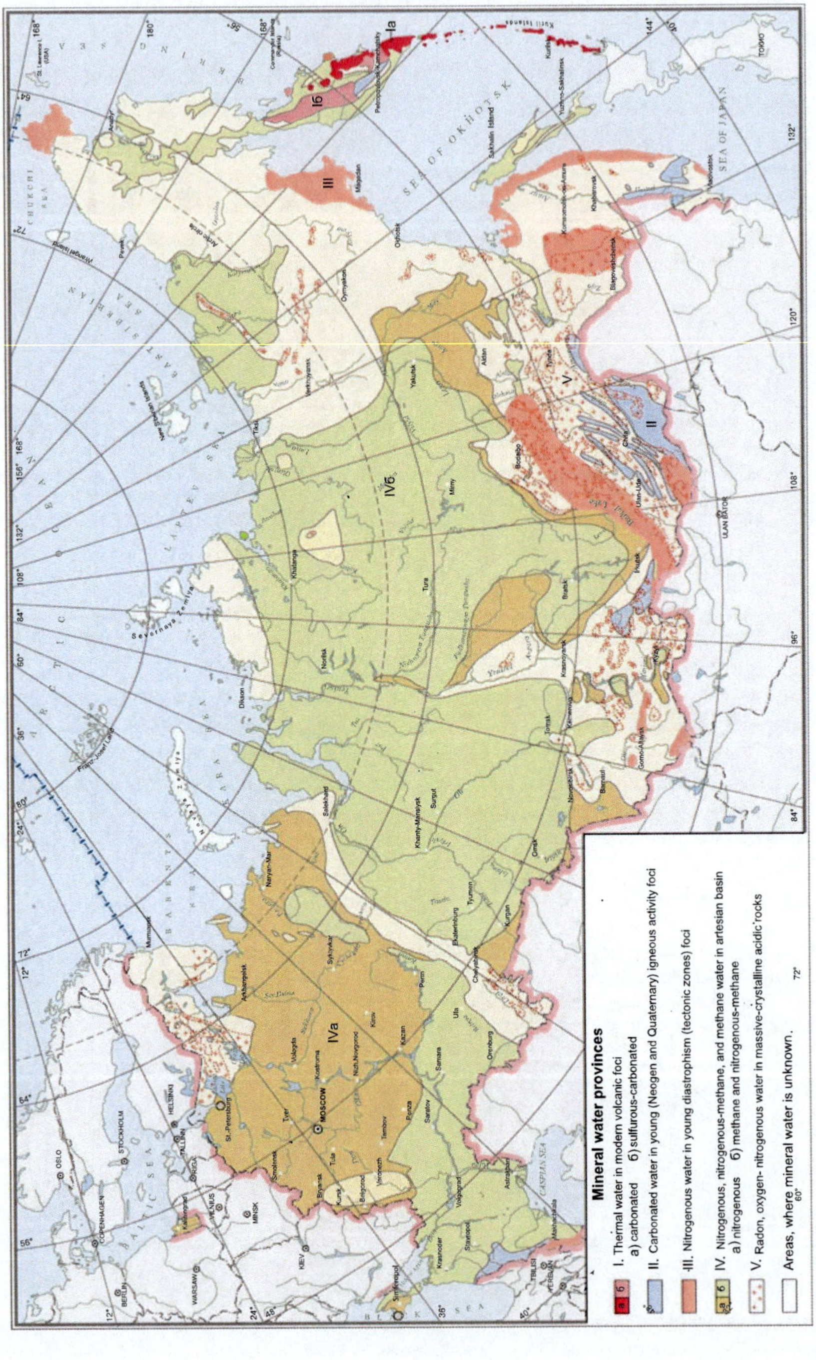

Map IV Mineral water provinces

regional information on healing springs is displayed. They include areas of international importance with a highly developed culture of medical services: Caucasus, Crimea, and promising remote areas (Tyva, Zabaykalsky Krai, Kamchatka) with great natural and healing potential, but so far poorly equipped and designed mostly for tourist groups. The special chapter is about the history of healing springs in Russia. The final chapter deals with health resort organizations, including the oldest Russian resorts using treatments based on natural water and mud deposits.

Reference

1. Ivanov VV (ed) (1968) Map of mineral healing waters of the USSR (1:4000000). Moscow (in Russian)

Chapter 1
Mineral Water

Abstract The chapter presents the classification of types of mineral waters based on their physical and chemical characteristics. Such parameters of mineral waters as mineralization, ionic and gas compositions, biologically active substances, temperature, and acidity (pH) are considered. Maps of the distribution of different types of mineral waters supplement the text. The therapeutic effect of mineral waters, including their external (balneological) and internal (drinking) applications is analyzed. The distribution and use of radioactive (radon) water for therapeutic purposes are examined. A specific section is devoted to the analysis of current health status of the population in Russia and the possibilities of healing effect of mineral water in those diseases where mineral water is in most common use.

Keywords Mineral water · Physical and chemical characteristics · Classification · Types of mineral waters · Distribution of mineral water springs · Radon water · Therapeutic effect · Balneological application · Health status · Russia

Water is the basis of life on the Earth. No known form of earthly life can exist without it. Water is a strongly polar universal solvent that can incorporate and retain many gaseous, mineral, and organic substances and compounds in various concentrations, thereby providing both living conditions and the internal homeostasis or metabolism of living organisms.

Water containing dissolved salts, trace elements, and other biologically active components is considered mineral water. Natural mineral water is a complex dispersed system containing a variety of minerals, organic additives, and gases. Moreover, it has been shown that absolutely chemically pure water (without impurities and substances dissolved in it) does not exist in nature (Gall 2015).

During the compilation of the electronic database of mineral water and preparation of the book, materials on 1075 springs located in various regions of Russia were generalized and analyzed (Malkhazova et al. 2019). The distribution of healing mineral water springs in the Russian Federation is shown in Map 1.1

There are many works on mineral waters and peloids (Jackson 1990; Guthrie, Mossman 1993 ; Reinbacher 1999; Van der Linden 2002; Van Tubergen, Varga

S. Malkhazova et al., *Healing Springs of Russia*,
https://doi.org/10.1007/978-3-030-83534-7_1

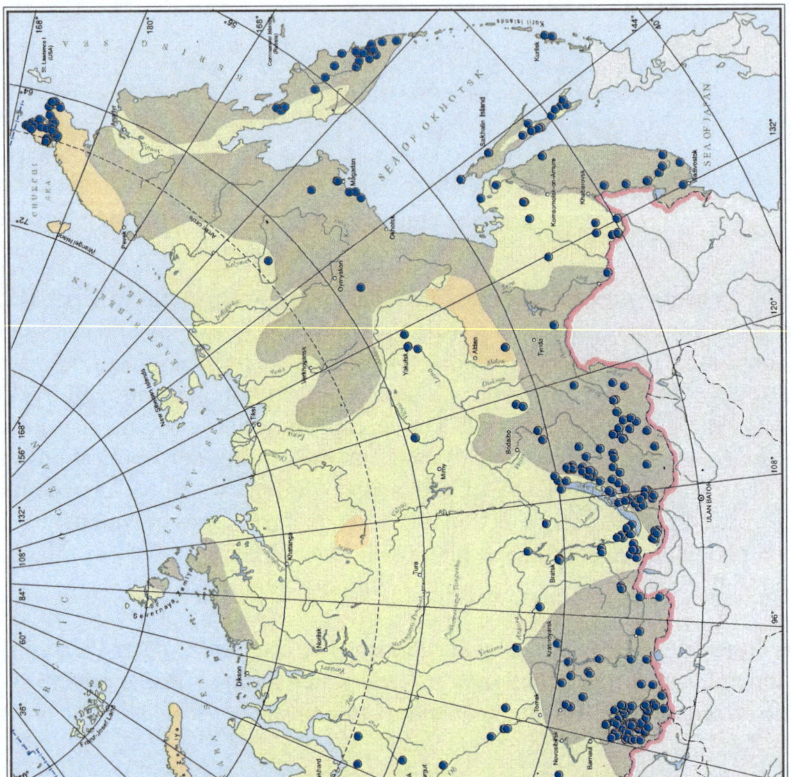

Map 1.1 Mineral water springs

2010). However, there are no such complex works and there are no works on such a large territory.

1.1 Types of Mineral Water Springs and Their Distribution

Mineral water includes natural water with chemical and physical properties (the content of various mineral and (less often) organic components, gases, radioactivity, acidic or alkaline reactions, etc.) that allow it to be used for therapeutic and preventive purposes.

Balneology (from lat. *Balneum*—bath, bathing) is a branch of resort medicine that studies mineral water and its therapeutic and preventive use, and takes into account the totality of biologically significant factors that characterize the effect of water on the human body. The main factors are the following:

1. general mineralization (M),
2. ion composition (anion/cation),
3. gas composition and gas saturation,
4. content of biologically active components (mineral and organic) in the water,
5. radioactivity (the content of dissolved radioactive components),
6. water temperature (°C), and
7. active reaction of water (pH value).

1.1.1 Mineral Water Classifications

Over the centuries, the diverse chemistry of natural water has inspired researchers to systematize them. A large number of classifications of water by its chemistry have been proposed. The earliest classification, used even before the development of assay analysis, was the division of water by predominant content of any salt, or its most characteristic feature. For example, there is sulfuric, ferruginous, arsenic water, etc. A similar division of water was made in the works of F. Hoffmann (1703), V.M. Severgin (1809), G.I. Hess (1825), and others (Shirokova 2010).

A whole series of works on mineral water classification appears in the USSR in the first half of the twentieth century:

1920—A.P. Gerasimov developed water classification by its gas composition and physical properties; the chemistry has subordinate importance in this classification.

1921—M.G. Kurlov proposed recording mineral water chemistry as pseudo-formula with anion content in the numerator and cation content in the denominator. Kurlov's advanced formula is used to date (Malkov et al. 2015), although it does not have independent classificatory significance.

1932—V.A. Aleksandrov developed a classification based on diverse indicators: composition of dissolved mineral components, nature of emitted gases, and presence of biologically active substances. Later, hydrogeologists declared that such a mixture of bases is unacceptable in the classification (Sulin 1946), but in medicine, this classification, improved by the author in 1956, is widely used (Mineral Water of the USSR—Ivanova 1974).

1933—V.I. Vernadsky, in his work on the systematical assessment of water's role on our planet, noted that natural water classification should develop in two directions. It is necessary to take into account the natural environment of water, on the one hand, and its chemistry, on the other hand (Vernadsky 1933).

1934—S.A. Zernov proposed hydrobiological classification of water types according to the mineralization level, dividing water into fresh, almost fresh, brackish, saline (marine), and brine. Subsequently, Knipovich (1938) scientifically substantiated the border between brackish and saline water. This division became the basis for the modern separation of mineral water by its mineralization level (Shirokova 2010).

1934—Based on the ion hydration theory, S.A. Shchukarev proposed classifying mineral water according to the six main ions present in water in more than 12.5% equivalents (as a proportional calculation of each ion concentration relative to the total equivalent concentration of minerals dissolved in water, taken as 100%). Considering that the main ions in natural mineral water are Na, Mg, Ca, Cl, SO_4, and HCO_3, a combination of three cations with three anions makes up 49 combinations that correspond to 49 classes of natural water (Shchukarev 1934). The hydrogencarbonate calcium, sodium chloride, and calcium sulfate classes were considered the main (primary) ones, because the water of these classes is formed as a result of direct rock leaching. All other classes are deemed secondary water formed as a result of metamorphization of primary (main) water.

1948—O.A. Alekin developed Shchukarev's classification with consideration of the ratio of equivalent mass that determines alkalinity and total hardness of water, and described the genesis of natural water according to the predominant anion. Cl-class (chloride) water has a predominantly marine origin: oceans, seas, estuaries, and relict lakes. C-class (carbonate, hydrogencarbonate) continental water makes up almost all rivers, fresh lakes, and groundwater. S-class (sulfate) metamorphosed water is genetically related to sedimentary rocks. The author considered it intermediate between C and Cl classes (Alekin 1948).

All subsequent natural water classifications can be considered variations of the systematics of S.A. Shchukarev and O.A. Alekin, because divisions of water mainly come down to a different form of expression of their composition, to the allocation of narrower or wider classes according to the assay of some component, or to different ways of distributing individual indicators in the classification headings.

In the USSR in the second half of the twentieth century, a hydrogeochemical mineral water classification was proposed, with consideration for its occurrence (Ovchinnikov 1963). It describes formation conditions in detail, as well as the composition of surface water (top water), groundwater, artesian, fractured, karst, and underground permafrost water, and defines mineral water corresponding to the above-described classifications.

In 1964, specialists from the Central Scientific Research Institute of Resort Medicine and Physiotherapy (Ivanov and Nevraev 1964) proposed an updated systematics that included seven main groups of mineral water: water without specific trace elements; weakly saline water; saline water and brines; carbonic water; ferruginous, arsenic and polymetallic water; siliceous thermal water; and radon water. Based on this classification, the State Geodesy and Cartography Administration of the USSR presented a variant of the territorial division of mineral water provinces, published in 1968 in the form of a large-scale map (see Map IV in the Introduction).

The main mineral water provinces in the USSR were the following:

I. Provinces of thermal water in the areas of recent volcanism and zones of active volcanism (Kamchatka Krai, Kuril Islands) and Quaternary volcanism (Caucasus Minor);
II. Provinces of carbonated water in the areas of young magmatic activity (Transcarpathia, Caucasus, including the region of Caucasian Mineral Waters, Eastern Sayans, Southern Primorye, Central Kamchatka, etc.);
III. Provinces of nitrogen thermal water in the areas of young tectonic movements—faults in the earth's crust (Tien Shan, Altai, Baikal region, the Far East);
IV. Provinces of nitrogen, nitrogen-methane, and methane water in the artesian basins occupying most of the territory of the USSR;
V. Provinces of radon, oxygen-nitrogenous water of the acidic crystalline rock massifs.

In medical practice, a modified combination of the Alekin and Alexandrov classifications is widely used. For a long time, it served as the basis of the USSR state standard "Mineral medicinal and medicinal-table drinking water." According to this classification, water is primarily divided by its anionic composition (according to Alekin 1948), and secondarily by the presence of biologically active substances and gases (according to Aleksandrov 1932). In total, there are seven mineral water classes distinguished by the predominant anions present in water in more than 12.5 equiv.%, whereby the part of other anions is less:

1. hydrogencarbonate (carbonate) alkaline water with predominant hydrogencarbonate anions (HCO_3),
2. chloride (salt) water with predominant chloride anions (Cl);
3. sulfate (gypsum) water with predominant sulfate anions (SO_4);
4. mixed water with complex anionic composition: hydrogencarbonate chloride, sulfate hydrogencarbonate, chloride sulfate, etc.; it contains more than two anions in an amount of 12.5 equiv.%;
5. biologically active water, including water of one of the four groups listed above, containing biologically active elements (iron, arsenic, fluorine, silica, iodine, bromine, boron, etc.);
6. gas water containing dissolved natural gas (carbon dioxide, sulfur, nitrogen, methane, etc.); gas water does not include artificially aerated water produced for bottling;
7. radon water containing radioactive gas radon (Rn).

Within each class, subclasses are distinguished by the predominant cation: sodium (Na), magnesium (Mg), calcium (Ca), or their combinations (sodium-calcium, magnesium-sodium, calcium-sodium-magnesium, etc.). The criterion for inclusion in the corresponding subclass is cation content in an amount of more than 5 equiv.%. Cations present in large quantities in the mineral water are indicated first.

Water hardness is an indicator directly associated with water mineralization, primarily with the content of Ca and Mg cations. There is a distinction made between total, temporary (carbonate), and constant (permanent) hardness. The total hardness is equal to the mathematical sum of the Ca and Mg cation concentrations in water. It includes temporary hardness, determined by the content of hydrogencarbonates (HCO_3) that combine when boiling with part of the Ca and Mg cations and form a dry residue (scale), and constant hardness associated with the presence of strong acid bases forming chlorides and sulfates.

In accordance with the international standard that went into effect in 2014 as the national standard of the Russian Federation (GOST (refers to a set of technical standards maintained by the Euro-Asian Council for Standardization, Metrology and Certification) 31.865–2012 "Water. Unit of hardness"), hardness is expressed in degrees of hardness (°H). 1°H corresponds to the content of Ca and Mg cations in water, numerically equal to 1/2 of their concentration in millimoles per liter (1°H = 1 mEq/l) (Table 1.1).

For practical purposes, for characterization of mineral composition of natural drinking water another state standard (GOST R 54.316–2011 "Mineral natural drinking water") is used. This standard is based on the classification that divides mineral water into extended groups depending on their chemistry, indicating both anionic and cationic composition. The groups distinguish hydrochemical types reflecting features that make it possible to give comparative characteristics and to indicate the similarity between water with a specific origin and any known analogue (type): Narzan-type water, Marcial type, Essentuki type No. 4 and No. 17, etc.

There are different lines of thought about natural mineral water classification: hydrogeologists study mineral water from the standpoint of the relationship between its origin and chemistry, medical and biological specialists and balneologists evaluate the effect of various mineral waters on the human body, and technologists work

Table 1.1 Groundwater classification by total hardness

Water groups	Hardness of water	
	Total concentration of Ca^{2+} *H* Mg^{2+}	Degrees hardness according to GOST 31.865–2012
Very soft	Less than 1.5 mEq/l	Less than 2°H
Soft	1.5–2 mEq/l	
Moderately strong	2–6 mEq/l	2–10°H
Strong	6–10 mEq/l	
Very strong	More than 10 mEq/l	More than 10°H

with standardization and water treatment. This difference in approaches to the classification of natural mineral water leads to difficulties in developing a unified scheme (Voronov and Vinograd 2004). In modern literature, a large number of terms and concepts are used for the classifications of mineral water because there is a need to take into account many varied parameters (mineralization, gas and ionic composition, acidity and alkalinity, temperature, radioactivity, presence of biologically significant trace elements, buffer capacity, hardness, origin, etc.). It indicates the complexity of developing a comprehensive classification (Kulikov and Voronina 2017).

In this regard, the systematic classification of water according to its geological and geographical features proposed in 1933 by the outstanding Russian natural scientist V.I. Vernadsky, is of particular interest. In his fundamental work "The History of the Minerals of the Earth's Crust" (Vernadsky 1933), he draws an analogy between water and minerals on the basis of atomic rather than ionic composition, and identifies large categories of natural water. Kingdoms are characterized by the main receptacles of water—surface (or land), underground, deep. The author distinguishes three kingdoms: (1) the surface water kingdom (surface water), (2) groundwater kingdom, and (3) deep water kingdom. The classification proposed by V.I. Vernadsky touches on the global state of water on our planet, including its vaporous, liquid, and solid states, and allows us to quantify the distribution of individual chemical elements in various families of natural water. Families are divided into species based on the chemistry, i.e., on the basis of chemical elements determined by assay of the water, regardless of the form in which they are found in it (in solution, in colloids, in living matter, in gases, etc.). The species is determined by the atoms prevailing in the water (except the hydrogen and oxygen atoms), in the order of their occurrence. In total, Vernadsky's classification covers 485 types of water; however, the author supposed that, in reality, their number could exceed 1,500. After dividing natural water into species, he indicated that there could be a further division into subspecies, etc., for individual, less common elements, for example, radium (Ra), which is the product of radioactive decay, or for other phenomena (number of bacteria, organic substances, etc.).

According to some modern authors, this classification can be considered the most ideal and promising for further development (Shirokova 2010), and its potential can be realized in the near future, taking into account the capabilities of modern science.

Today, modern physicists (Bulyenkov 2005; Konovalov 2013; Gall 2015) are very interested in water as the medium that provides living conditions on Earth from the point of view of both its chemical and physical qualities. Despite the fact that these basic studies are still experimental and require careful verification, for studying the functional structure of water there are already high-tech methods such as electron microscopy, magnetic resonance imaging, etc., being used, which make it possible to evaluate molecular and supramolecular water assay, as well as impurities present in it, even in minimal amounts, with a high degree of certainty (Semikhina 2005; Gall 2015). The results of today's experimental studies can be conceptual, changing humanity's concept not only of the importance of water for life support, but also about the physical nature of many biological phenomena, including the causes of diseases, processes of age-related change, and self-healing mechanisms of the body.

1.1.2 Characteristics of Natural Mineral Water

Almost all diverse classifications of natural mineral water that exist today are based on the scientific developments of specialists from previous generations. In constructing the maps in this book, the main principles of the approaches described here were taken into account, supplemented by a specially created recording system and graphic display of mineral water characteristics, used both for mapping and for creating an electronic database of healing mineral waters of Russia. By the beginning of 2019, this database included more than a thousand different mineral springs.

Mineralization. The degree of water mineralization depends on the concentration of mineral substances dissolved in it. According to the classification of mineral water by level of mineralization (Bogolyubov 2016) that is currently most common in Russia, the following degrees of mineralization of natural water (Table 1.2) are distinguished.

Fresh water, with salinity up to 600 mg/l, has the best taste. At the same time, according to organoleptic indications, WHO recommends an upper limit of mineralization of 1.000 mg/l for drinking water, i.e., to the lower limit of minor mineralization of water (Bogolyubov 2016). It should be noted that seawater, as the most common form of water on Earth, according to its degree of mineralization of 35 g/l, is located on the border of high mineralization water and brine mineral water according to this classification.

The balneological significance of the mineralization level is very high, because this indicator is one of the main criteria for classifying mineral water as water for internal or external use. So, for internal use, drinking water with a salinity of up to 15 g/l is used, and, along with this, it can be bottled and used for therapeutic purposes in regions remote from the place of the water's origin (GOST R 54.316–2011). External use of mineral water for baths and irrigation usually occurs near the spring, in the balneological resort. For external use, water with various mineralization levels can be used: from minor mineralized to strong brine. In some cases, for example, in the treatment of children, or if the concentration of aqueous brines is excessively high, diluted mineral water is used externally.

Table 1.2 Classification of mineral water by its mineralization level

Name	Mineralization
Fresh water	Less than 1 g/l
Minor mineralized water	1–2 g/l b
Low mineralized water	2–5 g/l
Medium mineralized water	5–15 g/l
High mineralized water	15–35 g/l
Brine mineral water	35–150 g/l
Strong brine water	More than 150 g/l

The mineralization level distribution of mineral water springs considered in the book and mineralization data for it are presented in Fig. 1.1, and their geography is shown on Map 1.2

Ionic composition. To characterize the properties of mineral water in Russian literature, a water chemistry formula proposed by Kurlov and Sobkevich (1921) at the beginning of the past century is widely used. It is a pseudo-fraction displaying the basic properties and assay of chemical elements dissolved in mineral water. Anions are indicated in the numerator of the fraction, cations are indicated in the denominator. Along with this, it takes into account the relative content of ions with more than 5% of their total mass dissolved in a liter of water; its value is called the equivalence percentage (equiv.%). Next to the ion symbol, its content in equiv.% is indicated, which makes it possible to compare the ionic composition of water with different mineralization. At the beginning of the formula, the content of dissolved gases (CO_2, H_2S, N_2, CH_4, O_2, etc.) and the mineralization level (g/l) are usually given, and the second part includes the presence of other active elements (Fe, B, Br, I, As, Si, etc.), pH level, water temperature (T), and (if relevant data are presented) flow rate of the spring (Q). Compilation of the water chemistry formula is a prerequisite for the chemical analysis of mineral water for conformance to the GOST R 54.316–2011.

This is an example of the water chemistry formula for Marcial mineral water (Gerasimenko et al. 2016):

$$\mathrm{Fe}\,64\,\mathrm{Mr}/\pi\mathrm{M}\,0.7\,\frac{\mathrm{SO}_4\,97}{\mathrm{Mg}_{40}\mathrm{Ca}_{25}(\mathrm{Na}+\underline{\mathrm{K}})_{22}}\mathrm{pH}\,4.0\,\mathrm{T}\,8.0\,^{\circ}\mathrm{C}$$

This formula is read as follows: ferruginous (total iron is 64 mg/l), fresh (total mineralization is 0.7 g/l), sulfate (SO_4 anions in the numerator are 97 equiv.%), magnesium-calcium-sodium (in the denominator, respectively: Mg, 40 equiv.%; Ca—25 equiv.%; Na and K—22 equiv.%), acidic (pH—4.0) cold water (temperature on emergence from the spring is 8.0 °C).

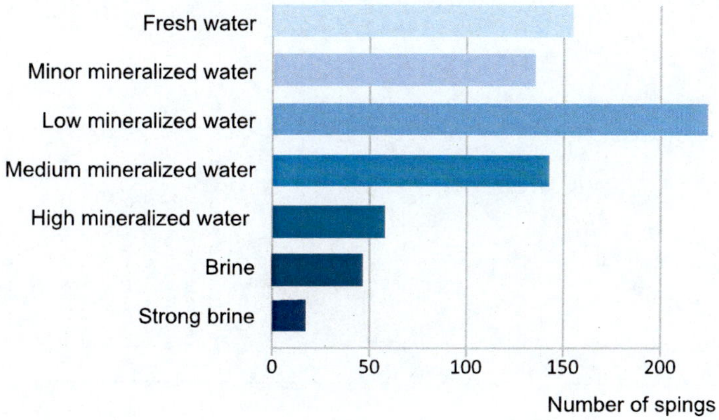

Fig. 1.1 Distribution of mineral water by its mineralization level

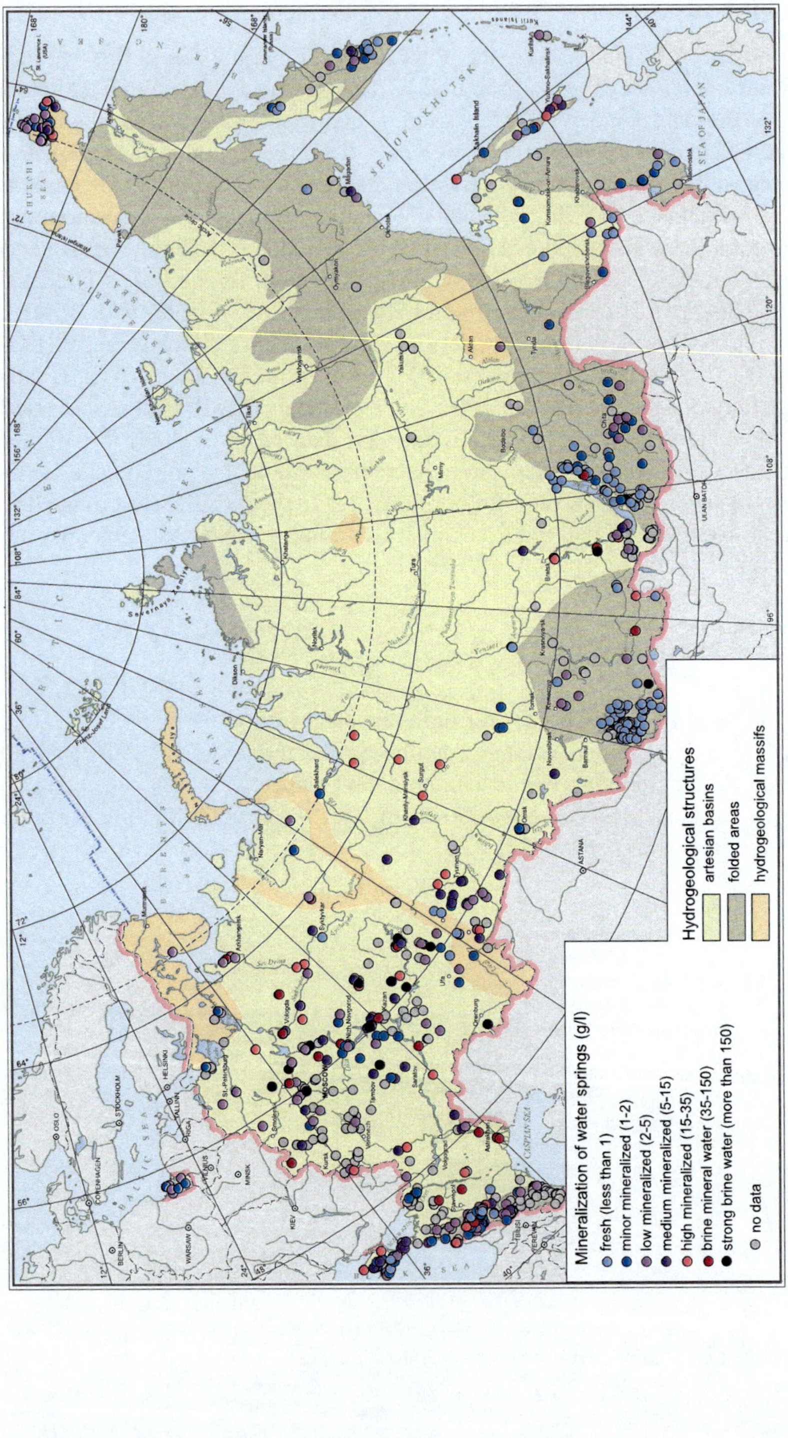

Map 1.2 Water mineralization

Taking into account the existing classifications and characteristics (Kurlov and Sobkevich 1921; Aleksandrov 1932; Shchukarev 1934; Alekin 1948; Ovchinnikov 1963; Ivanov, Nevraev 1964; Ivanova 1974; GOST R 54.316–2011, etc.), an original method of displaying the ionic composition and basic physical and chemical properties of mineral water by colored ring nomograms has been developed and used (Starikov 2018).

Visualization of the chemistry ratio as colored ring nomograms displaying percentage of cations and anions in water as sectors with different areas and colors and indicating its mineralization level, acidity, radiation activity, or other parameters in the central part of the nomogram, allows us to evaluate visually the chemistry and other characteristics of water (Figs. 1.2 and 1.3).

This method of displaying information can be used both for characterization of an individual water and for the cartographic representation of various springs in a given territory. It makes it possible to evaluate the chemistry, quantity, and geographical location of mineral water springs in a mapped area clearly, informatively, and efficiently.

To conduct a comparative analysis of data from various springs (especially those where the quantitative characteristics of water are not fully reflected or there is only comparative characterization of their hydrochemical types), a system of twin nomograms was used, displaying subgroups of the anionic (−) and cationic (+) composition in the form of two adjacent hemispheres of various colors or textures (Fig. 1.4).

Considering that the ionic composition of mineral water is very diverse in this large amounts of information, it is advisable to distinguish the main subgroups by combining the same types of water containing anion or cation in sufficient quantities for classification in such a way that it is convenient to record and process statistical material without using bulky pseudo-formulas that do not readily lend themselves to

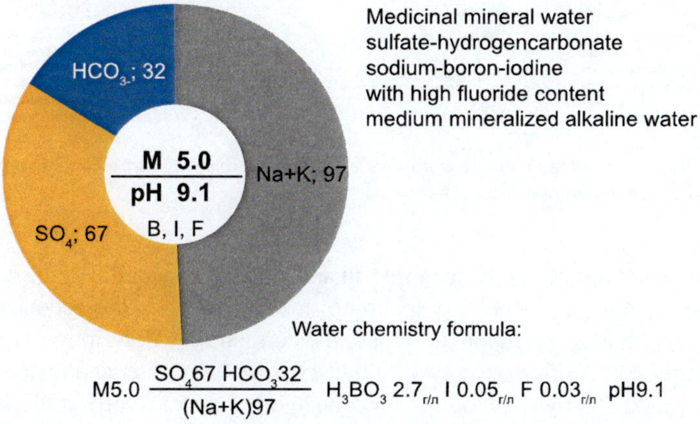

Fig. 1.2 Ring nomogram of Lazarevskaya medicinal mineral water. Krasnodar Krai, Volkonsky deposit, well 84-E (according to GOST R 54.316–2011)

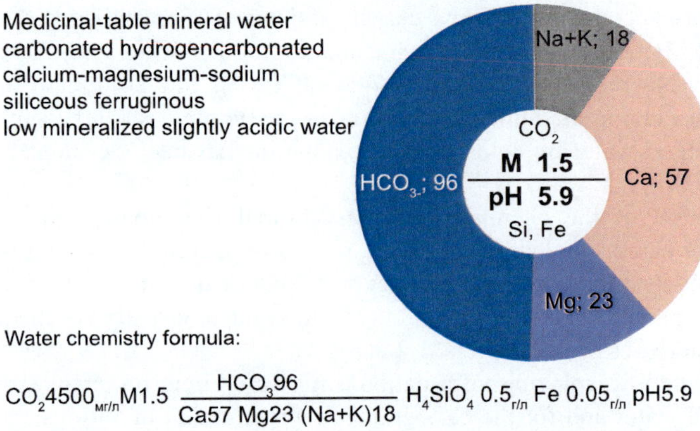

Fig. 1.3 Ring nomogram of medical-table mineral water Shmakovskaya, Primorsky Krai, Shmakovka deposit, wells 2-E, 4-E, 5-E (according to GOST R 54.316–2011)

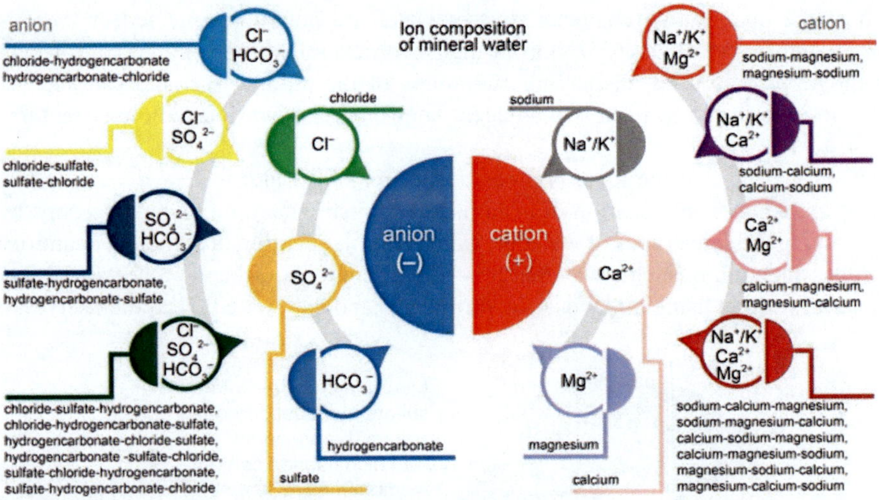

Fig. 1.4 Scheme of anionic/cationic composition and principles of formation of mineral water name depending on its predominant ions

digital analysis. This allows the meaningful and concise transmittal of the descriptive textual part of the ionic water composition in a digital code corresponding to the subgroup, and makes it possible to bypass the comparative description of a specific mineral water that is often used even in the scientific literature, and instead to use a reference to a well-known analogue called the hydrochemical type (GOST R 54.316–2011), for example, Essentuki-type water or Narzan-type water.

This principle of mineral water classification by its anionic/cationic composition does not exclude additional information about chemical component assay, but it makes it possible to highlight the most significant ions, combining them into the corresponding subgroups.

As noted previously, the main groups of mineral waters, divided by their anionic composition, are chloride, hydrogencarbonate, and sulfate water. Chloride waters prevail in the underground hydrosphere and occupy most of the sedimentary cover. They are spread over the entire area of Russia and usually occur in marine sedimentary deposits at a depth of 100–150 m on the plains and 250–350 m in more widely traversed and elevated places. In areas with saline rocks, chloride water lies close to the surface. The total thickness of the chloride water zone reaches 2–3 km, in places, it is up to 4–5 km (Ovchinnikov 1963).

Usually, sodium chloride predominates among chloride water, less often there is chloride calcium-sodium and sodium-magnesium-calcium water, which is of the greatest interest for balneology.

Mineralization of chloride water varies in a very wide range: from minor and medium to brine. Calcium chloride and calcium-magnesium brines usually contain high concentrations of bromine and form an independent balneological group. Chloride water is more often classified according to gas composition as nitrogen and nitrogen-methane. Chloride water stimulates metabolic processes in the human body, improves the secretion of the gut, pancreas, and small intestine. They are used for the treatment of gastrointestinal disorders, neuropathies, and circulatory disorders (Bogolyubov 2016).

Natural hydrogencarbonate water is characterized by a high content of hydrogencarbonate anion because of the interaction of water with limestone (Ovchinnikov 1963). Hydrogencarbonates in natural water are primarily calcium and magnesium salts; they exist only in a dissolved state, and upon heating and boiling they turn into insoluble carbonates. Calcium in hydrogencarbonate water promotes its anti-inflammatory effect, and magnesium does the same for its antispasmodic effect. That is taken into account in the treatment of patients with respiratory and gastrointestinal diseases (especially inflammatory ones with a spasmodic tendency). This water is widely used in a variety of drinking (table, therapeutic-table, medicinal) and balneotherapy water. Water with sodium cation (hydrogencarbonate sodium water) alkalizes the gastric content. Depending on the method of application, they can both stimulate and inhibit gastric juice secretion, therefore they are considered universal for various gastrointestinal pathologies (Kipiani 1974). Due to the decrease in acidosis, they promote carbohydrate metabolism normalization. It is used for metabolic disorders and diabetes mellitus (Frolkov and Mikhaylyuk 2014). In general, this water, due to its detoxifying effect, helps restore the acid–base balance in the human body. Therewith, a change in the blood alkaline reserve affects the reabsorption of fluid in the renal tubules, so hydrogencarbonate water is used in urolithiasis treatment (Kulakov and Oransky 1970).

Sulfate and sulfate–chloride water are formed in the upper zone of the sedimentary cover in rocks containing gypsum or anhydride, dissolution, and leaching of which leads to sulfate accumulation in the water (Voronina and Zelinskaya 2009). The cationic composition depends on the degree of washing of water-bearing deposits: in well-washed rocks there is calcium accumulation, in less washed rocks there are calcium and magnesium ones. In submerged structures, sulfate–chloride sodium water is more often formed.

Sulfate water stimulates gastrointestinal motility, and especially beneficially affects the restoration of liver and gall bladder functions. It is used for biliary tract diseases, chronic hepatitis, diabetes mellitus, and obesity (Klyachkin and Schegolkov 2000; Frolkov and Mikhailuk 2014). Such water is widespread in the European territory of Russia and in the southern part of Eastern Siberia. For therapeutic external and internal use, sulfate–chloride and sulfate-hydrogencarbonate water are also often used.

The distribution of the mineral water springs considered in the book by their anionic/cationic composition is shown in Fig. 1.5, and their geography is shown in Map 1.3, 1.4, 1.5, 1.6, 1.7, 1.8 and 1.9.

The group of largest distribution is *chloride water* (Map 1.3). Within it, chloride sodium water predominates significantly: it represents more than 16% of all mineral springs and is the most common natural mineral water in the Russian Federation. The second place in chloride water is occupied by sodium-calcium springs, which are found in no more than 3% of cases in total. Sodium-calcium-magnesium water is even less common (Fig. 4.1.6).

In second place among the sources of mineral water with a simple anionic structure is *hydrogencarbonate water*, which makes up more than 15% in total (Map 1.4). Its chemistry is highly diverse. Within this water, there are calcium-magnesium, sodium-calcium, sodium-calcium-magnesium, and calcium water types in comparable amounts. In this group, the amount of sodium water is significantly less than in the previous one.

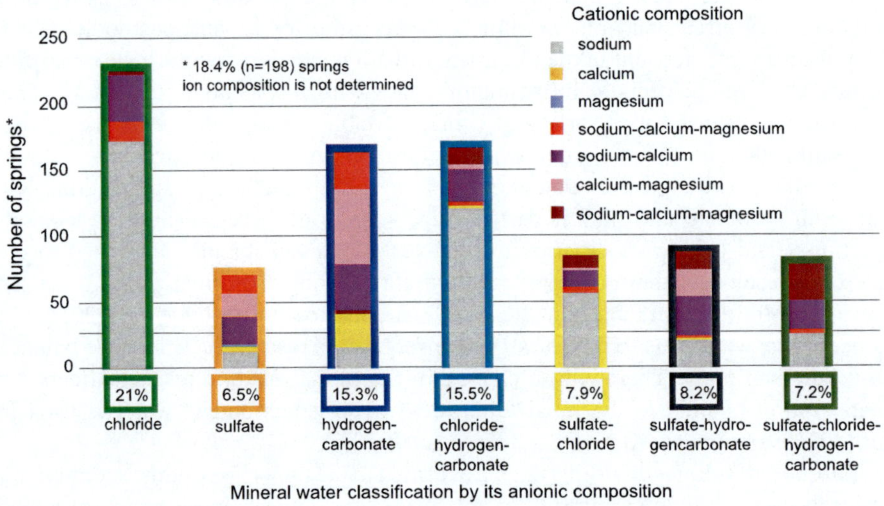

Fig. 1.5 Distribution of water springs by the anionic/cationic composition

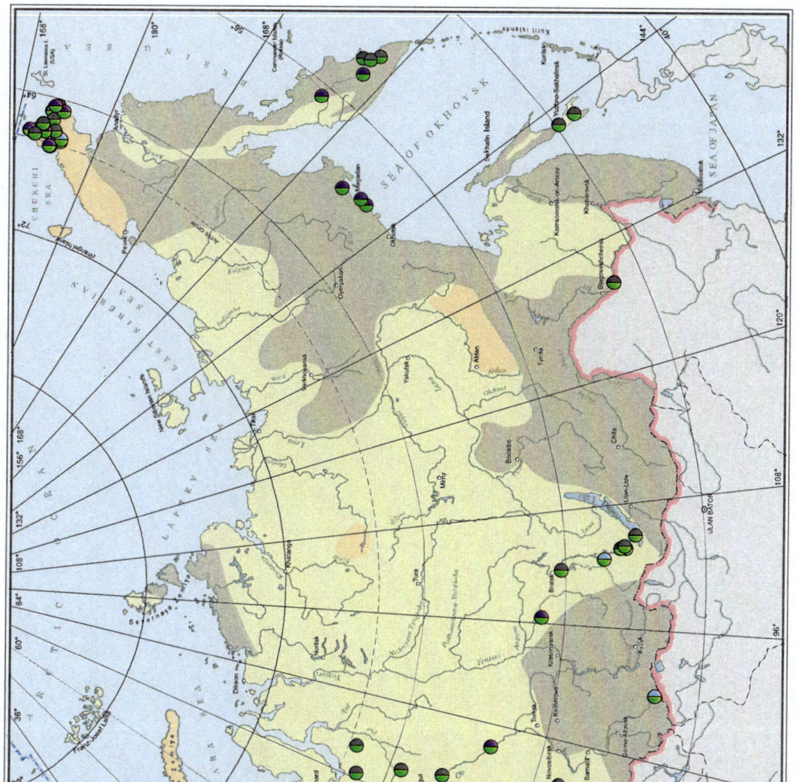

Map 1.3 Chloride mineral water

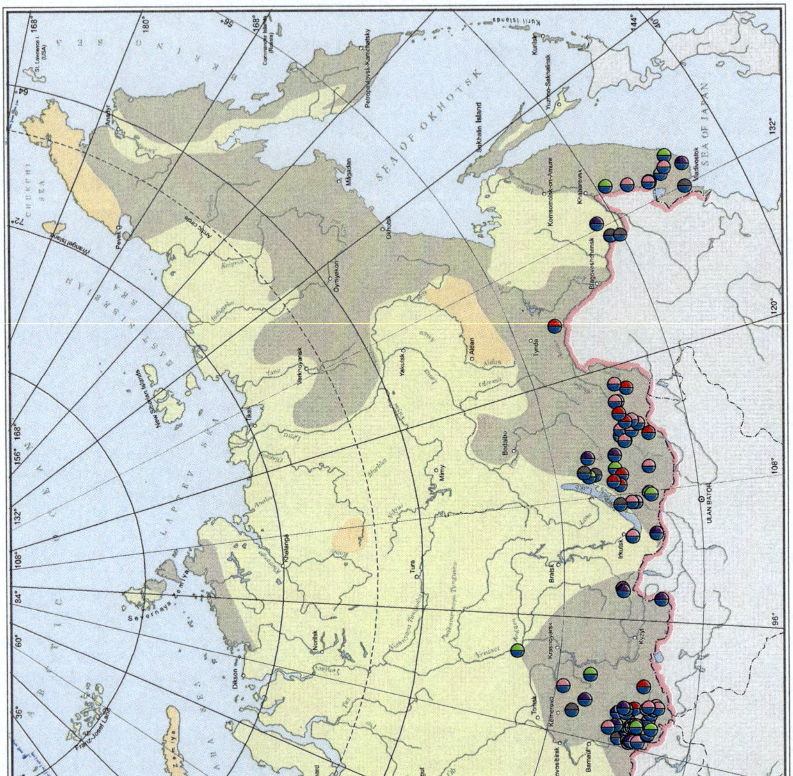

Map 1.4 Hydrogencarbonate mineral water

Sulfate water (Map 1.5) is much less common than the first two groups. In total, it makes up only 6.5% of all mineral springs, but the cationic composition of its internal structure is very diverse and includes almost all types of cationic groups: sodium, calcium, magnesium, sodium-calcium, calcium-magnesium, and sodium-calcium magnesium.

Other mineral water springs (which, according to our data, are slightly less than 40%) may be classified as mixed minerals by their ionic composition. Their distribution in Russian territory is shown on the Maps (1.6, 1.7, 1.8, and 1.9). Water with different ions in its chemistry in an amount of more than 5 equiv.%, is usually classified as mixed water. The name of this water is formed in the order of the predominant anions and cations; in water with insignificant differences in ion concentrations, this order is not of fundamental importance.

The ions in mixed mineral water affect the water structure in different ways depending on their hydration (Samoilov 1957; Gall 2015). A distinction is made between positive and negative hydration. Positive hydration decreases the movement of water molecules near the ion. They become less mobile than in pure water. Negative hydration increases the mobility of water molecules near the ion. Ions are divided by hydration into those: (1) ordering the water structure around itself (Cr, Zn, Mg, Li, Na, CO_3, SO_4, OH, F, Cd); and (2) loosening the water structure around itself (Cs, K, Cl). The degree of ion hydration is one of the main factors determining their mobility and chemical activity (Voronina and Zelinskaya 2009). The presence of a large number of electrolytes and organic substances in mineral water significantly changes its physicochemical properties. These changes affect the healing effect of mineral waters with different ionic chemistry. Mixed mineral water with complex mineral composition has a combined effect on the human body. Metabolic disorders, gastrointestinal, circulatory, genitourinary, and musculoskeletal diseases can be indications for the use of mixed mineral water (Frolkov and Mikhailyuk 2014; Badalov et al. 2016).

Chloride-hydrogencarbonate water is the most widespread (Map 1.6) among mixed ones. It represents more than 15% of the total number of mineral springs. Noteworthy is the similarity of the internal distribution structure of chloride-hydrogencarbonate and chloride water, with sodium water in a leading position. In this case, the share of chloride-hydrogencarbonate sodium water is more than 11% of all mineral springs, which makes them second in prevalence within the Russian Federation after chloride sodium water (16%).

Two-thirds of sulfate–chloride springs (Map 1.7) are also sodium springs. This can be explained by the genetic affinity of chlorine anions and sodium cations that are historically the main ions of the ancient seas. At the same time, in the structure of sulfate-hydrogencarbonate water (Map 1.8) and sulfate–chloride-hydrogencarbonate water (Map 1.9), no such pronounced cationic advantage is observed.

It should be noted that 18.4% (Map 1.10) is water with nondetermined ionic composition, because no reliable information about the anionic and cationic chemistry of its water has been received from the literature or from responses to the Russian Geographical Society requests in the regions. On the one hand, this may be due to technical difficulties in determining the ionic composition of water locally because

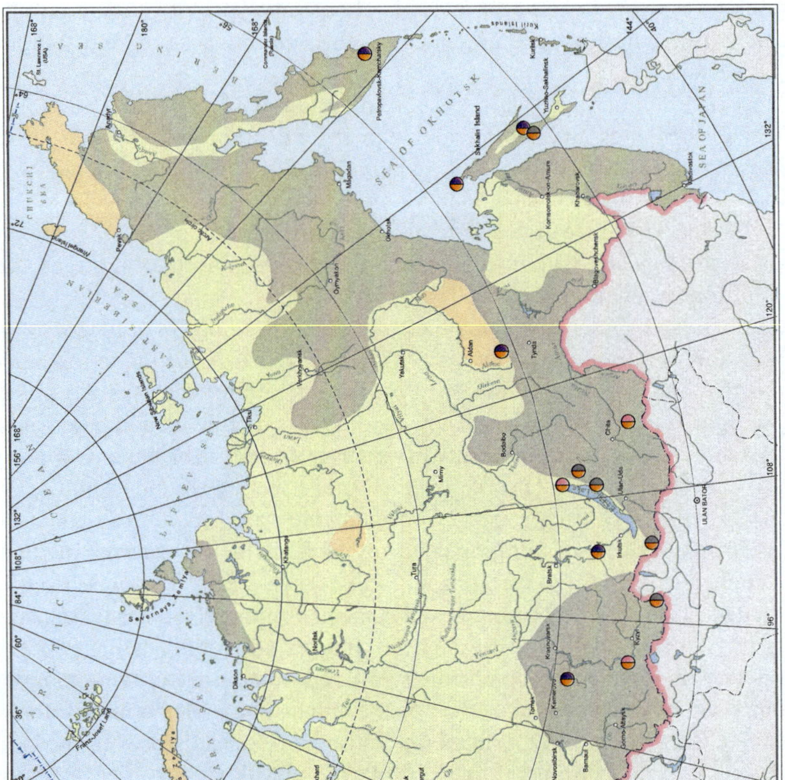

Map 1.5 Sulfate mineral water

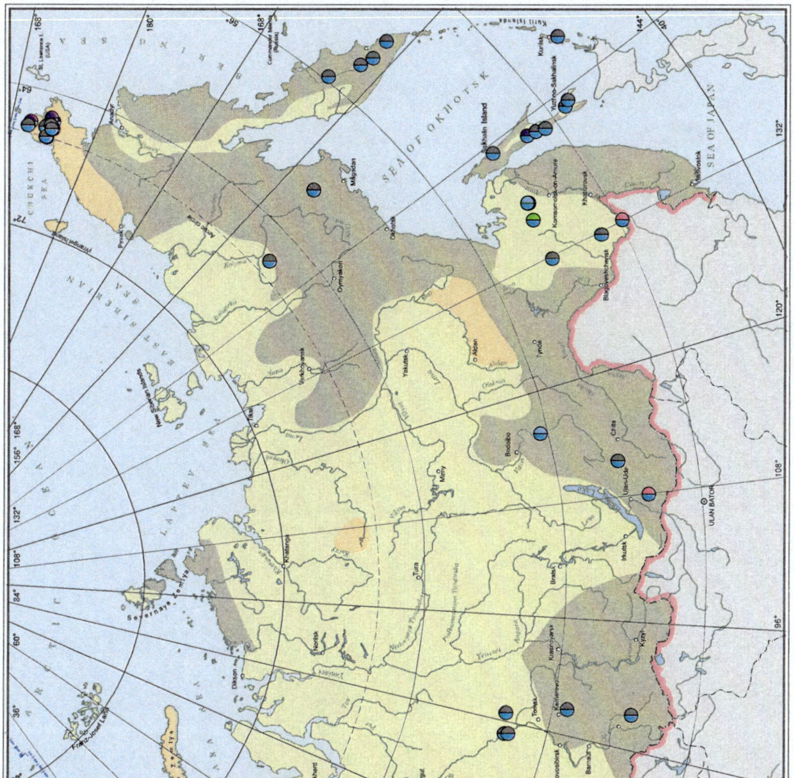

Map 1.6 Chloride-hydrogencarbonate mineral water

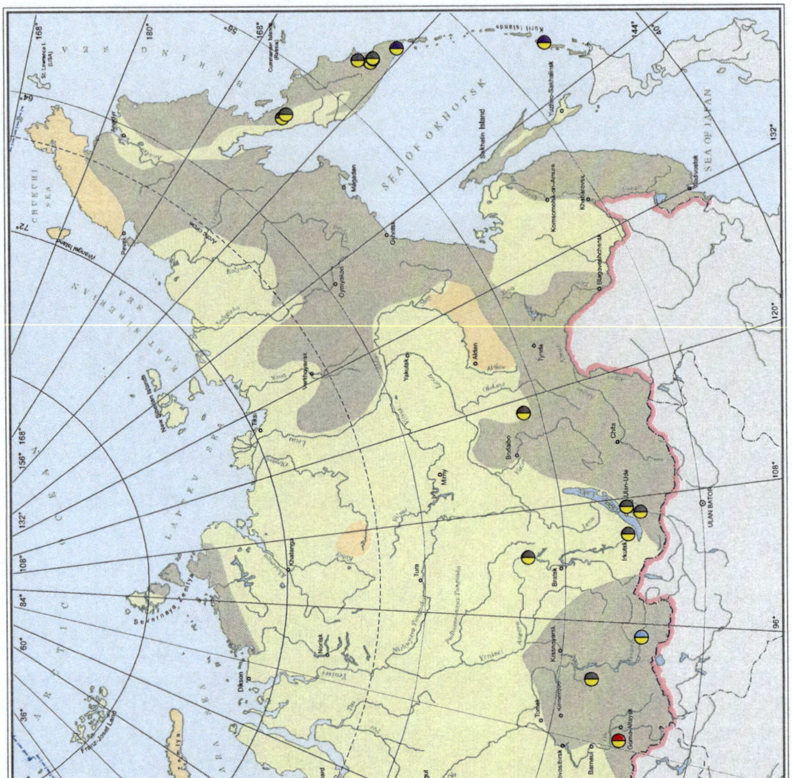

Map 1.7 Sulfate–chloride mineral water

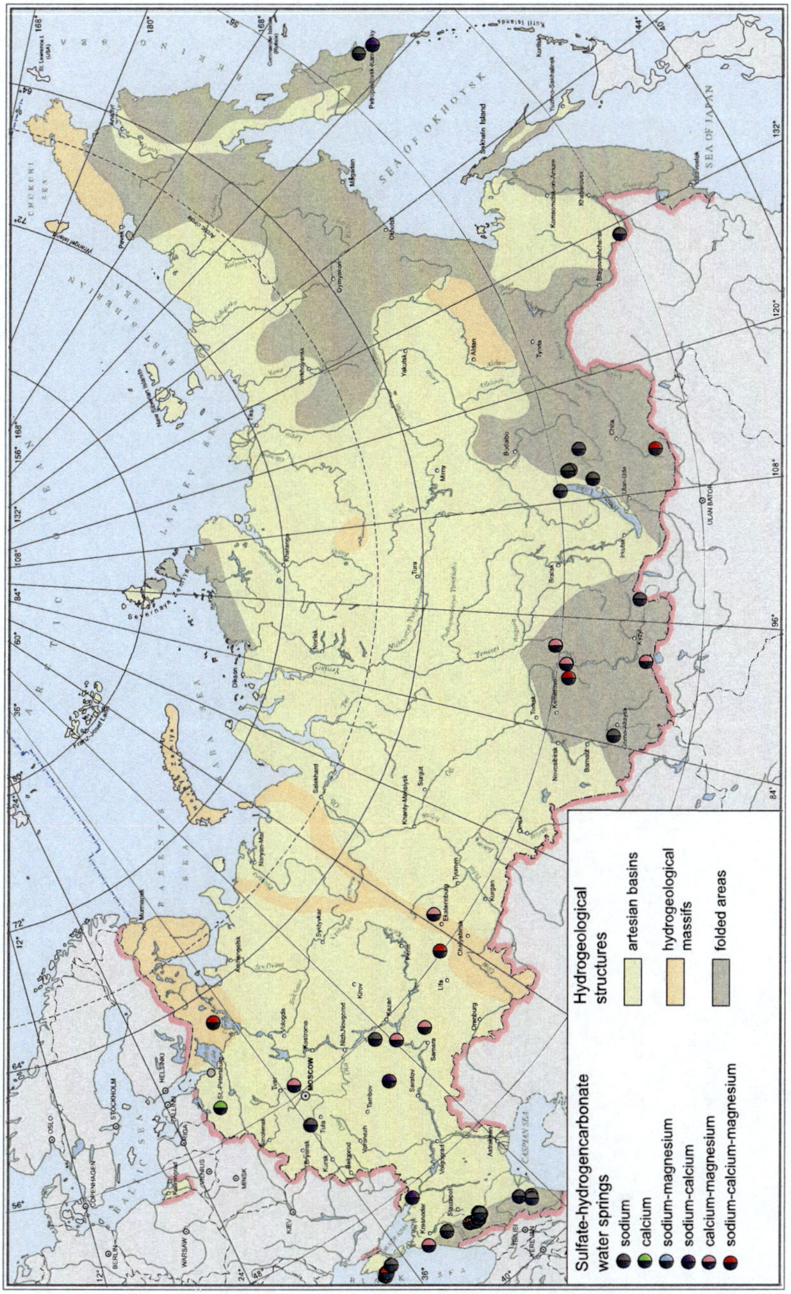

Map 1.8 Sulfate-hydrogencarbonate mineral water

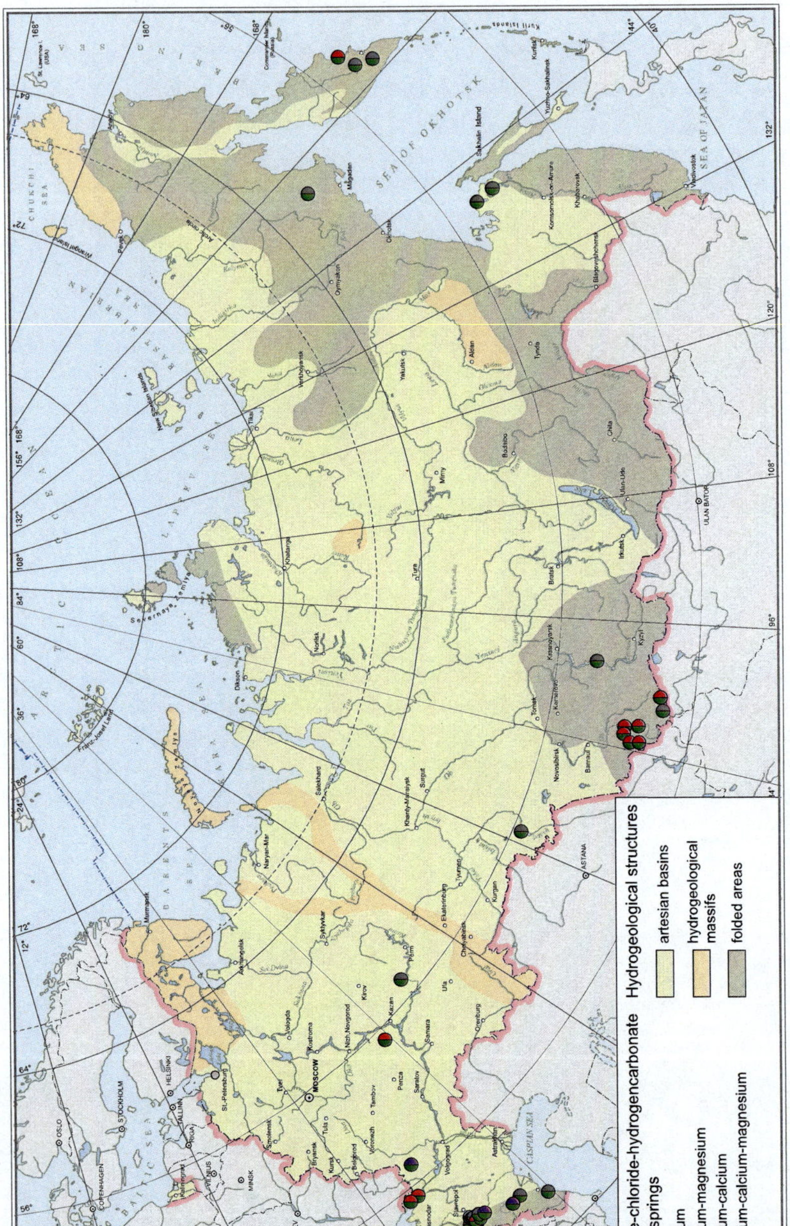

Map 1.9 Sulfate–chloride–hydrogencarbonate mineral water

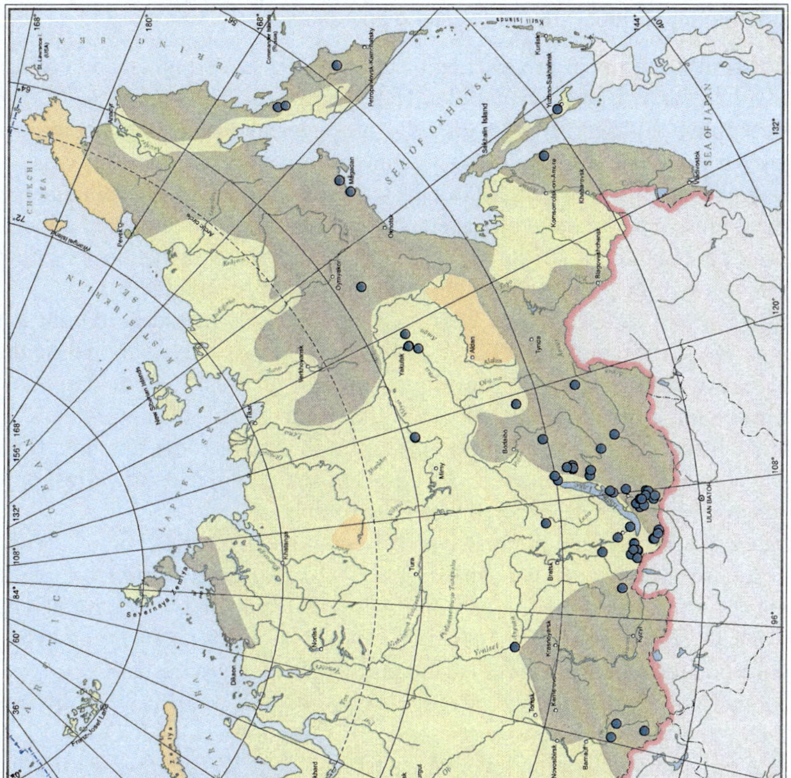

Map 1.10 Nondetermined ionic composition water

there is a lack of specialized laboratories in many regions, and, on the other hand, due to the imperfection of the classifications used and the presence of many outdated data from the Soviet period, difficult to record and analyze.

Gas composition. In terms of the gas composition, water is distinguished between carbon dioxide (CO_2), hydrogen sulfide (H_2S), nitrogen (N_2), methane (CH_4), and more complex (nitrogen-carbon dioxide, nitrogen-methane, carbon dioxide-methane, etc.) water types. When classifying mineral water according to gas composition, gases representing more than 10% of the total volume of all gases (spontaneous and dissolved) present in the water are taken into account. The distribution of mineral springs by gas composition is shown in Fig. 1.6. It should be noted that the exact gas composition of a significant part of water springs is currently undefined.

Carbonated water in accordance with modern standards include groundwater containing more than 500 mg/l (GOST R 54.316–2011) of dissolved carbon dioxide. The main areas of carbonated water development are Caucasus, Central Asia, East Sayan, Transbaikal, Primorye, Sakhalin, Central Kamchatka (Ivanova 1974). In tectonic terms, these areas are structures of the alpine system or zones of epiplatform orogenesis. Outside these provinces, in the more ancient geological structures, only rare local manifestations of carbonated water are found.

The content of the main biologically active component of this group—carbon dioxide—varies over a wide range: from 0.5–3 to 20–40 g/l in the deep horizons of the largest deposits (structure of the Caucasian Mineral Water), which is determined by an increase in CO_2 solubility at high pressure. Carbon dioxide is found in groundwater in the dissolved and spontaneous phases, the ratio between phases is 1:4.

The chemistry and general mineralization of carbonated water is very diverse: HCO_3–Ca (Mg); SO_4–Ca; HCO_3–Na; SO_4–Na; HCO_3–Cl–Na; Cl–HCO_3–Na; Cl–Na; Cl–Na–Ca; its mineralization level varies from 1–5 to 15–35 g/l or more. The

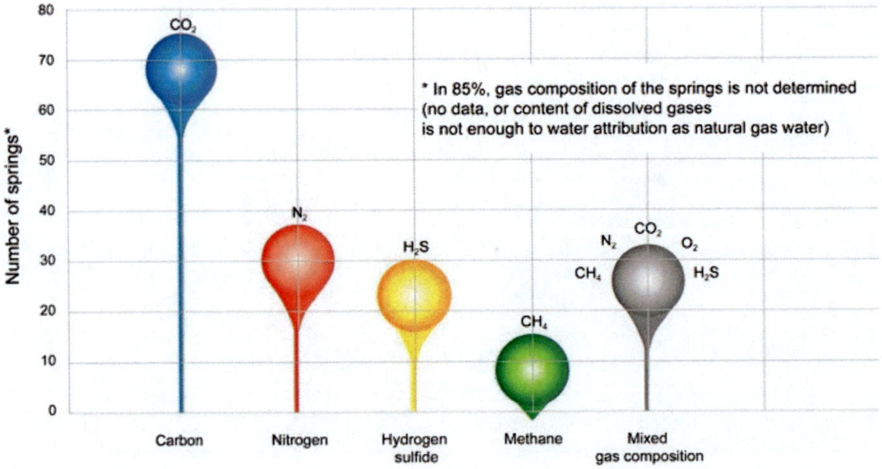

Fig. 1.6 Distribution of the mineral springs by their gas composition

formation of the chemistry is associated with a complex combination of several leading processes interacting in the rocks—carbon dioxide—water system. It should be borne in mind that the main components of the aqueous phase of carbonated water can be infiltration (atmospheric) water and sea (ancient or modern) water, the mixing and geochemical transformation of which in the earth's crust give rise to all main hydrochemical types (Ovchinnikov 1963).

A large role in the formation of water chemistry is played by rocks undergoing dissolution and leaching in the presence of carbon dioxide. So, for example, the most mineralized carbonated water (mineralization of more than 35 g/l) is formed during the dissolution of halogen formations. In accordance with structural-tectonic zonality, the general regional zonality of carbonated water also changes appropriately (Ivanov 1968). In the most ancient and, consequently, more washed-out structures of epiplatform orogenesis areas (Transbaikal, Primorye), the most developed are HCO_3–Na and HCO_3–Ca low-mineralized water types (Darasun). In areas of young alpine folding (Caucasus, Kamchatka Krai), HCO_3–Cl–Na, Cl–HCO_3–Na, and Cl–Na types are widely represented; they are the products of the transformation of water of marine genesis.

An important feature of carbonated mineral water is the presence of a wide range of microelements, including biologically active ones: silicic acid, arsenic, iron, boron, and some others (lithium, cesium, rubidium, mercury, fluorine, germanium). The assay of these elements and their state in certain types of water is determined by the geochemical features of the structures and the chemistry of water: the greatest number of them are noted in the HCO_3–CI–Na and CI–HCO_3–N types of carbonated water.

Hydrogen sulfide (sulfurous) water is very diverse in terms of formation, and therefore it is represented by different types in ion composition, H_2S assay, and mineralization. Concentrations of hydrogen sulfide water vary from minor and low mineralized to super-high brines with a salt content of more than 600 g/l.

The origin of sulfide water is very diverse—from ancient buried brine and marine sedimentation water to infiltration leach water. The most widespread water is the water of mixed (atmospheric and marine) and atmospheric genesis (Ovchinnikov 1963).

The vast majority of sulfurous water contains biogenic hydrogen sulfide formed as a result of sulfate reduction by sulfate-reducing bacteria using organic matter from bedrock (Kulikov et al. 1991). There is also hydrogen sulfide of chemical (including thermochemical) genesis buried in sediments of biogenic origin, but it is less common. In the presence of sulfates and organic matter, biogenic sulfate reduction processes encompass thick sedimentary strata, and sulfide water forms regionally concentrated aquifer systems (or a multistory reservoir system, as in oil and gas provinces) with a high content of hydrogen sulfide.

Nitrogenous mineral water. Nitrogen is the main component of atmospheric air, but it is poorly soluble in water under ordinary conditions: only 2.5 parts of nitrogen can be dissolved in 100 volumes of water. Therefore, the assay of dissolved nitrogen in springs in provinces of nitrogenous mineral water, which are widespread in country (Ivanova 1974), is insignificant. At the same time, nitrogen contained in the depths of the earth's crust, at high pressure and temperature, can accumulate in water up to 20–25 mg/l and come to the surface as part of nitrogen-siliceous thermal springs. The water of these springs usually has low mineralization (up to 2 g/l) and various ion compositions. It is a naturally warm and hot alkaline water containing free nitrogen and a large amount of silicic acid (50–150 mg/l). Such water can be formed as a result of atmospheric water penetration of tectonic cracks in deep (up to 2–3 km) horizons of the earth's crust and following leaching of crystalline and volcanic-sedimentary rocks (Voronina and Zelinskaya 2009).

For medicinal purposes, nitrogenous water is used in the form of general and local baths, bathing in pools, showers, irrigation, intestinal rinsing, inhalation, and less often for a drink for treatment. Procedures using nitrogenous mineral water have a painkilling and sedative effect, stimulate blood circulation, and promote endocrine and metabolic normalization. Nitrogenous water is used for musculoskeletal diseases, neuropathies, cardiovascular disorders, skin diseases, gynecological diseases, and some endocrine disorders (Badalov et al. 2016; Bogolyubov 2016). High-temperature nitrogenous mineral water (70–90 °C) in some cold areas is used to heat residential buildings, greenhouses, etc.

Methane mineral water has limited application for medicinal purposes. The therapeutic effect of methane water is due to the complex effect of its biologically active substances on the human body. For the use of this water, balneotechnical devices have to be made that isolate methane escaping from the bath in a special gas vent and then remove it into the atmosphere. The external use of methane mineral water is effective for cardiovascular, skin, and other diseases.

Water containing biologically active substances. Biologically active mineral water also includes mineral water containing various microelements in biologically active concentrations. Primarily, this is ferruginous water, bromine and iodine-bromine water, and siliceous, arsenic-containing, and boron-containing mineral water. Table 1.3 shows mineral water containing biologically active substances, which is classified as drinking water in accordance with GOST R 54.316–2011.

Table 1.3 Drinking mineral water containing biologically active substances

Name of group of mineral water	Name of biologically active component	Value of the mass concentration biologically active component	
		Medicinal, Mg/I	Medicinal-table, Mg/I
Ferrous	Iron (total)	–	No less than 10.0
Arsenic	Arsenium	No less than 0.7	–
Boron	Boracium (expressed as orthoboric acid)	No less than 60.0	35.0–60.0
Siliceous	Silica (expressed as metasilicic acid)	–	No less than 50.0
Bromine	Bromine	No less than 25.0	–
Iodine	Iodine	No less than 10.0	5.0–10.0

Ferruginous mineral water is water with a total iron content (Fe^{2+} and Fe^{3+}) of more than 10 mg/l. Iron enters the groundwater due to the destruction of rocks and minerals (mainly sulfides) by the infiltration of water into certain geostructural zones of the earth's crust. In terms of gas composition, the most common is ferruginous nitrogen and carbonated water.

The most famous ferruginous water in Russia is Marcial water, discovered even during the reign of Peter the Great. Its main value is the high concentration of active (divalent) iron, which reaches 95 mg/l (much more than in other ferruginous springs in Russia and the world). However, this feature of Marcial water makes it possible to store it only in dark containers and only for a few hours. Due to the rapid oxidation of iron, its biological activity drops significantly, which makes it inappropriate to transport the water for use outside the resort.

Carbonated water has the greatest distribution among all groups of ferruginous water. Increased iron content in carbonated water is observed in almost all main provinces. This is due to the general geochemical conditions of its formation: the source of iron in carbonated water is host rocks, during leaching in the presence of carbon dioxide, bivalent compounds are formed, mainly iron hydrogencarbonates with high solubility in a wide pH range from 3.5 to 8.0.

In carbonated water, iron is a typomorphic component in the following hydrochemical types:

1. hydrogencarbonate magnesium-calcium and sodium-magnesium-calcium minor mineralized water (1–3 g/l);
2. sulfate-hydrogencarbonate (hydrogencarbonate-sulfate) magnesium-calcium sodium water with low mineralization (2–5 g/l);
3. chloride-hydrogencarbonate (hydrogencarbonate-chloride) sodium and calcium-sodium water with low mineralization (3–7 g/l).

Nitrogenous ferruginous water is formed in zones of orefield oxidation due to the intense leaching of iron by acidic (pH 1.5–3.0) mine water. It is characterized mainly by sulfate or chloride-sulfate chemistry with a complex combination of Fe, AI, Mn, Zn in the cationic part, and sometimes Na (Manshina 2007).

Another group of nitrogenous ferruginous waters is genetically related to the underground water of loose quaternary sediments, which is characterized by a diverse cationic composition. The anionic composition of this group includes mainly sulfates and hydrocarbons, the mineralization level corresponds to fresh water (0.3–0.7 g/l); the iron content does not exceed 0.07 g/l (Polyustrovsky water).

Relatively rare, increased concentrations of iron are observed in hydrogencarbonate sodium water (Borjomi type) and in highly mineralized (15–35 g/l) sodium chloride water, sometimes reaching brine concentration. Ferruginous water is very valuable medicinal and medicinal-table drinking water (Arshan, Batalinskaya, Darasun, Kuka, Lastochka, Marcial, Polyustrovo, Malka, Shmakovka, Elbrus, etc.). There are large resort complexes based on their springs.

The therapeutic use of *bromine, iodine, and iodine-bromine water* is determined by the following threshold values: for bromine more than 25 mg/l, and for iodine more than 10 mg/l. Bromine usually is an iodine satellite; iodine water with a low assay of bromine is relatively rare. Since the accumulation of high concentrations of bromine in groundwater is mainly associated with the concentration of sea water, it is predominantly of sodium chloride, calcium-sodium, or calcium brine composition. The last type is widely developed in the deep horizons of the eastern half of the European part of Russia and Eastern Siberia. The total amount of bromine depends on the salinity of the sedimentary stratum of the artesian basin and ranges from a few mg/l to 9–10 g/l. Dissolved gases are mainly nitrogen and methane; hydrogen sulfide and carbon dioxide are sometimes present.

High iodine concentrations are found both in brines and in less mineralized water (Azov, Aniva, Semigorsk No. 6, Sochi, Talitsa, Khadyzhensk, etc.). In terms of chemical composition, it is chloride sodium and hydrogencarbonate-chloride sodium water; in terms of gas composition, it is mainly methane water, in some cases with hydrogen sulfide and sometimes carbon dioxide.

Siliceous waters, or siliceous terms, include an extensive group of water with a silicic acid content of more than 50 mg/l and a temperature of more than 35 °C (Fig. 1.7). The main siliceous drinking water types are Amur, Arshan, Bykogorsk, Gaaz, Kuka, Novotersky, Pyatigorsk No. 1, Tersinskaya, Shadrinskaya, Shmakovka, etc. This group is most widely represented among carbonated water and nitrogen terms formed during the leaching of silicates of acidic igneous rocks. Siliceous thermal waters are distributed mainly in the areas of manifestation of recent and modern tectonic and volcanic processes.

Nitrogenous siliceous thermal water is the most representative group, characterized by the following main features: low mineralization (up to 0.5 g/l and 0.5–1.0

Fig. 1.7 The output of nitrogenous siliceous natural terms with temperature from + 30 to + 95 °C on the banks of the Klyuchevka river, Malka, Kamchatka Krai

g/l), sharply alkaline reaction (pH from 8.5 to 9.5), high temperature (35–50 °C to 100–200 °C in the deep parts of deposits), high concentrations of fluorine and radon.

Carbonated siliceous terms are also quite widespread, mainly within the mountain-folded regions (Northern Caucasus, Primorsky Krai, Kamchatka Krai). Compared to nitrogenous terms, this group has a higher mineralization (3–7 g/l, rarely up to 10–15 g/l), conditional carbon dioxide content (0.9–1.1 g/l) and temperatures up to 50–75 °C.

Methane siliceous terms are relatively rare: they have very different chemistry and mineralization—from minor mineralized (4–5 g/l) to highly mineralized (more than 15 g/l).

The distribution of silicic acid (the main biologically active component) is clearly controlled by the temperature regime of mineral water. A concentration of up to 50 mg/l is usually observed in low-thermal (20–35 °C) waters; with an increase in temperature to 50 °C, its content increases up to 80–100 mg/l. The maximum concentrations of silicic acid (up to 150 mg/l) are specific to high-thermal water. Siliceous terms are very valuable medicinal mineral water of both drinking and balneological profiles.

The group of *arsenic-containing (arsenic) water* includes groundwater with an arsenic concentration of more than 0.7 mg/l. Arsenic accumulates in groundwater due to a number of geochemical processes. The main processes are the following: (1) leaching from rocks at high temperatures and pressures in the presence of carbon dioxide; (2) supply of arsenic from deep endogenous (magmatic) solutions; (3) extraction of arsenic by acid mine water during the destruction of minerals (realgar, auripigment) in hydrothermal deposits of arsenic ores.

By degree of enrichment with arsenic, the following types of underground mineral water are distinguished:

1. weak—As content from 0.7 to 5.0 mg/l,
2. strong—As content from 5.0 to 10.0 mg/l,
3. very strong—As content more than 10.0 mg/l.

Arsenic-containing carbonated water is characterized by a diverse ionic composition: from sodium hydrogencarbonate and calcium-sodium to sodium chloride, with a mineralization range of 3–5 to 20–35 g/l. Most carbonic arsenic-containing water is classified as weak by its arsenic concentration, a much smaller part is strong, and only a few deposits are very strong. Among the specific components in paragenetic association with arsenic in carbonic waters, boron, metasilicic acid, and iron (Vinokurov 2008) can be noted.

Acidic mine-type water with a high content of metals (Fe, AI, Cu, Mn, etc.) differs from the previous ones both in composition and formation conditions. In the gas composition of this water, nitrogen is 86% v/v and oxygen 9.5% v/v. It is formed in the ore oxidation zone, which determines its low temperature (2–19 °C). Acidic fumarole-type arsenic thermal water is hydrogen sulfide carbonated by its gas composition. In terms of ionic composition, it is sulfate, less often chloride with complex cationic composition. Its mineralization does not exceed 5.0 g/l. Arsenic concentrations in this water do not exceed 1.2–1.8 mg/l. Siliceous arsenic-containing terms are related to nitrogenous-carbonated and carbonated types of water. Their total mineralization does not exceed 2.0 g/l. This water has very high temperatures (88–100 °C), which sharply reduce CO_2 solubility. Its arsenic concentration is 1.0–1.5 g/l. For drinking use, Sinegorie and Chvizhepse arsenic mineral waters are mainly used.

Boron-containing water, which includes groundwater with an orthoboric acid concentration of at least 50 mg/l, is very widespread (Yevpatoriya, Essentuki, Zaramag, Lazarevsky, Malka, Mukhena, Izhevsk, Sakhalin, Semigorsk, Sinegorsk, Chizhepse, etc.). It is known that boron has a pronounced biological, pharmacological, and toxic effect. It can affect the metabolism, the functional state of neuroregulatory processes, and has antimicrobial properties. But an overdose of it can cause boric toxicosis. Boron-containing water is used in the treatment of gastrointestinal diseases, metabolic disorders, vegetovascular dysfunctions, weakened immunity, and chronic fatigue syndrome.

Radioactive (radon) water. Radioactive (radon) water contains radon gas (Rn) and radium (Ra) in high amounts. Radon enters the water from rocks due to the decay of the radium contained in them.

The main importance in radioactive water formation is the chemistry and structure of rocks. The most intensive process of radium and radon entry into water occurs in highly fragmented (or fractured) rocks at a relatively low flow rate of water (Ovchinnikov 1963).

For therapeutic purposes, radon water with a radioactivity of more than five nanocurie (nCi) is used. According to balneotherapeutic signs, two subgroups of radioactive mineral waters are distinguished: subgroup A includes radon water with simple chemistry in which radon is the only therapeutic component; subgroup B is radon water of complex chemistry, where, in addition to radon, there are other biologically active components or properties, such as metasilicic acid, carbon dioxide, high water temperature, etc.

The major part of radon water is composed of mineral water of the first subgroup: there is cold oxygen–nitrogen minor mineralized water of a shallow circulation of atmospheric origin; the radon assay in it can reach hundreds of nCi/l.

Radon water of the second subgroup is mainly distributed in the mountain-folded regions of Caucasus and Eastern Siberia (Transbaikal, Altai). Nitrogenous siliceous alkaline thermal water with relatively low activity (up to 20 nCi/l). is most widely represented in it. From a balneological point of view, very important is carbonated minor- and medium-mineralized cold and subthermal water with radon concentrations from 10–40 to 100–200 nCiL, which has a wide range of therapeutic effects (Pyatigorsk, Molokovka, etc.).

In Russia, the most famous springs of radon water are in the Urals: Lipovsk radon water (low radon concentration—Rn 5–40 nCi/l) and Uvildinsk radon water field (high radon concentration—Rn > 120 nCi/l). In Bashkiria, there are springs of radon water (low radon concentration—Rn 20 nCi/l) at the Krasnousolsk resort (Fig. 1.8). In Western Siberia, there are well-known springs of radon water: Belokurikha (Rn 10–12 nCi/l); Zaeltsovsky (Rn 10–12 nCi/l); in Eastern Siberia are Nilova Pustyn (Buryatia)

Fig. 1.8 Krasnousolsky radon springs

and Tumnin (Krasnoyarsk Krai); in the Far East are Kuldur (Jewish Autonomous Oblast) and Nachiki (Kamchatka Krai).

The Radon Clinic in Pyatigorsk is the largest radon balneological complex. Up to 2,500 procedures of 17 different types can be undertaken there per shift. The Pyatigorsk radon water field is characterized by varied radon content and chemistry of water: high radon (Rn 230–330 nCi/l) water of the Beshtaugorsk deposit, medium radon water (Rn 70–80 nCi/l) with complex ionic composition, and low radon (Rn 15–40 nCi/l) heat sulfur springs (Smirnov-Kamensky 1971).

Mineral water temperature. Temperature is the main parameter of the thermo-physical properties of mineral water. It determines the solubility and content of gases in water and specifies the therapeutic effect of chemicals dissolved in water. Water obtained from different depths has different temperatures (Ovchinnikov 1963). Depending on the temperature regime, mineral water can have opposite effects. Cold water stimulates the gastric motion function. Warm mineral water relieves spasms well (Vorobiev and Parfenov 1982).

The temperature of mineral water can be very different, ranging from highly chilled water (below 0 °C) in permafrost areas to hot water (over 100 °C) in areas of volcanoes and geysers. Superheated waters can also be opened by boreholes in the deep aquifers of artesian basins and outside volcanic regions. However, waters are not used in medicine to treat people if they are too cold or too hot (Fig. 1.9). Medicinal mineral water is classified into categories according to temperature (°C) (Ovchinnikov 1963; Ivanov, Nevraev 1964; Kulikov et al. 1991) (Fig. 1.10):

Fig. 1.9 Thermal village of Yelizovo district of Kamchatka Krai is part of the Paratunka resort complex. Heating of houses in the village is carried out by thermal water

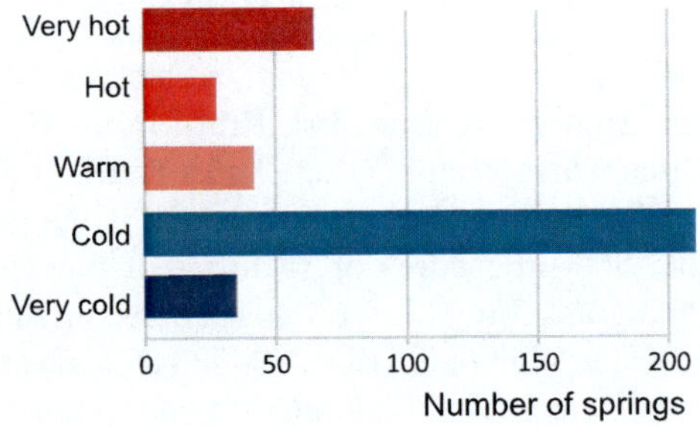

Fig. 1.10 Distribution of mineral water by its temperature

1. very cold (0–4),
2. cold (4–20),
3. warm (subthermal) (20–37),
4. hot (thermal) (37–42),
5. very hot (hyperthermal) (42–100).

In resort bathing practice, the temperature is usually not allowed to be higher than +38 °C. Hyperthermal water has to be cooled before its use. Hot water is of great medicinal value, since it does not require heating.

The temperature of mineral water depends on the thermal regime in the depth of its circulation. Supercooled water with a temperature of less than 0 °C is common in areas of permafrost (the north and northeast of the Russian Federation). Very cold water is common in Siberia and other areas of permafrost, and cold water is common in the middle latitudes. Warm water (subthermal, weakly thermal, akrotherm) is distributed at depths from several hundred meters to 1.5 km. Hot water (thermal, homothermal—close to the temperature of the human body) is distributed in young mountain-folded areas at depths of several hundred meters, and in platform at depths of 1.5–2 km. Very hot (hyperthermal) water (Fig. 1.11) is found in young mountain-folded areas with deep tectonic disturbances (Map 1.11). Exceedingly hot water (superheated water of steam hydrotherms and geysers) is widespread in areas of modern active volcanism. It can be used in the energy industry (Ovchinnikov 1963; Ivanov and Nevraev 1964; Kulikov et al. 1991).

There is a definite relationship between temperature, gas, and the chemistry of water. Most of the cold water types are calcium, and most of the warm and hot water types are sodium. The hottest water is often saturated with nitrogen, and there are many cold waters among the carbonated. Hydrogen sulfide water is both cold and warm, less often hot. Ferruginous water of the lowlands is usually cold. There is a relationship between water temperature and the radon content in it. The higher the temperature of the water, the greater the amount of radon transferred from solution to gas.

Mineral water acidity (pH). The acidic properties of water are determined by the action of hydrogen ions, and alkaline by hydroxyl ions. The activity of hydrogen ions is quantified by the pH value. This value in mineral water can vary from 0.4–0.5 to 8.5–9.5. Most water types are characterized by a low content of hydrogen ions. The concentration of hydrogen can be very high only in water containing various acids. Acid–base conditions determine the rates of element migration and the relationship between the transition processes of chemical elements into mineral water and their deposition from water. Mineral water is divided into six groups according to its pH (Ovchinnikov 1963; Ivanov, Neraev 1964; Kulikov et al. 1991) (Fig. 1.12):

1. strongly acidic (less than 3.5),
2. acidic (3.5–5.5),
3. slightly acidic (5.5–6.8),
4. neutral (6.8–7.2),
5. slightly alkaline (7.2–8.5), and
6. alkaline (more than 8.5).

The distribution of mineral water by pH in the Russian Federation is presented on Map 1.12 The thermal water of modern volcanism areas is characterized by the lowest pH values (pH less than 2). For water from the sulfide ore oxidation zones and highly mineralized water of halogen formations of large platform structures, an acidic reaction is typical (pH of 2 to 5). The thermal water of crystalline massifs (siliceous terms), as well as mineral water of marginal troughs and intermountain depressions are usually characterized by an alkaline reaction. Mine water of alkaline massifs is sharply alkaline. For many types of mineral water, especially sodium hydrogencarbonate, there are typical temporary changes in pH, which are associated with the conversion of carbonate equilibria (Kulikov et al. 1991). Despite the fact that the pH reaction is not the basis for classifying a water as mineral, it to some extent reflects the effect of water on the human body.

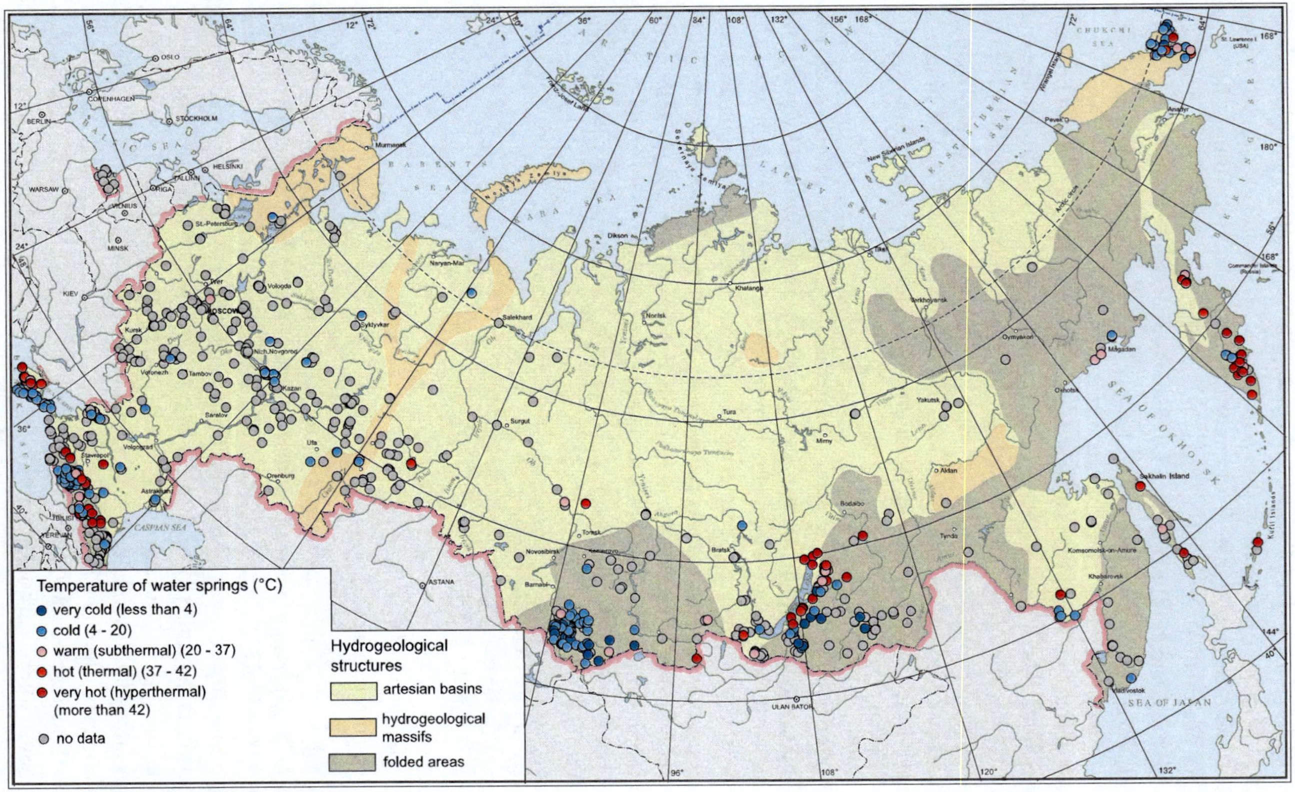

Map 1.11 Mineral water temperature

Fig. 1.11 Hyperthermal (93 °C) mineral water output to the surface. Kamchatka Krai

Fig. 1.12 Distribution of mineral water by its pH value

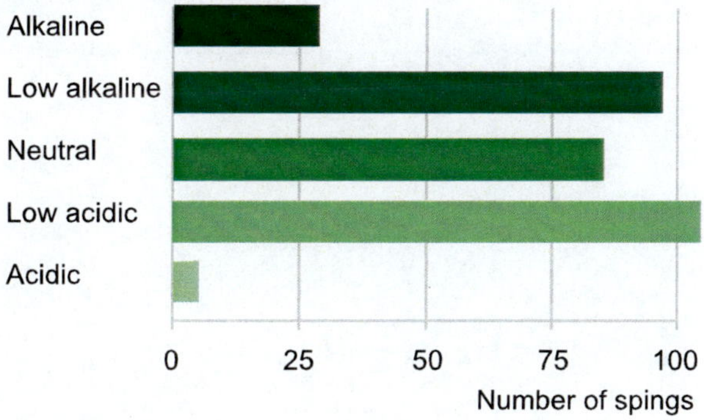

1.2 The Use of Natural Mineral Water for Treatment

The healing effect of mineral water depends on its physicochemical properties and methods of its application. It is evaluated clinically based on a study of the objective changes that occur in the human body under the influence of water. The most pronounced effect is observed with combined (external and internal) exposure to the human body. However, some mineral water can't be used for internal (drinking) use (for example, brines and water with a high level of radioactivity). At the same time, the external use of certain mineral water is not always advisable due to its unproven effectiveness (for example, low-saline water) or for economic reasons (high production cost or limited flow rate). Fig. 1.13 presents the distribution of water springs considered by the nature of their application.

As can be seen from the graph, mineral water is most often used for therapeutic purposes in a combination of ingestion and external use. The high proportion of

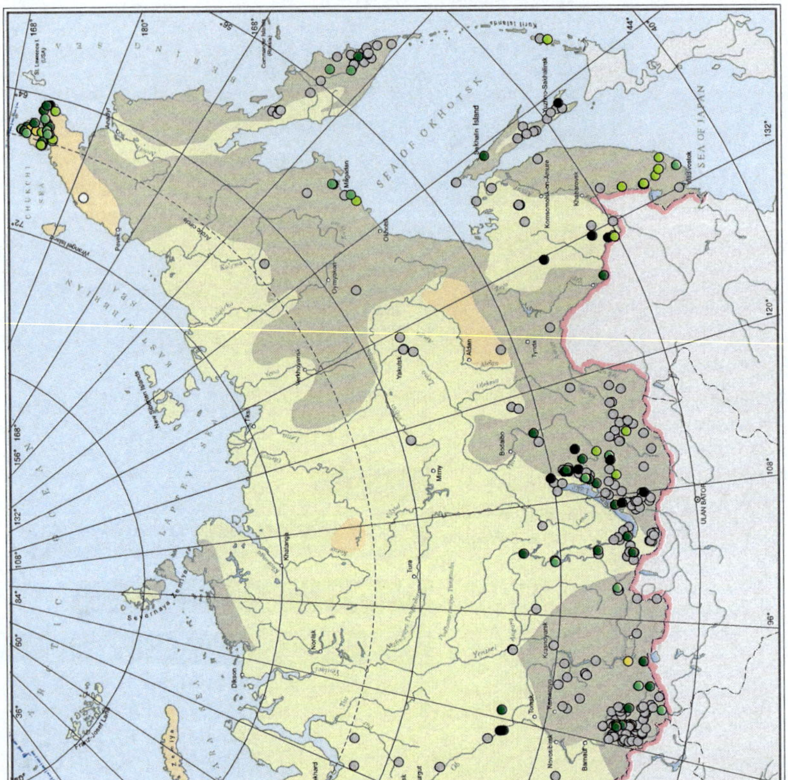

Map 1.12 Mineral water acidity

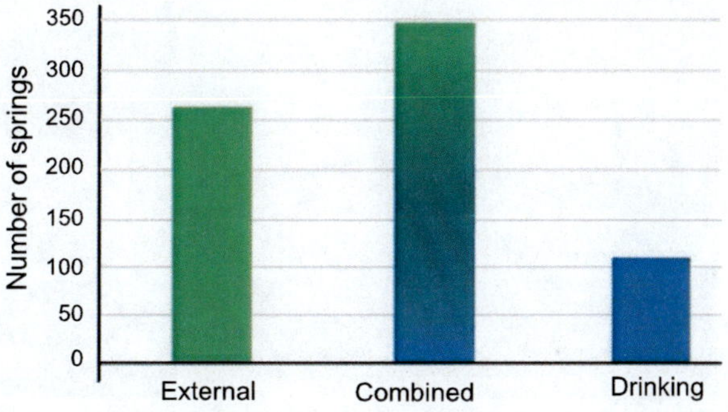

Fig. 1.13 Distribution of water springs considered by nature of their application

exclusively external use in relation to exclusively drinking use is most likely due to the inclusion only of data on local drinking mineral springs in this sampling, excluding the use of the water for bottling and sale in other regions.

1.2.1 Drinking Use of Mineral Water

According to GOST R 54.316–2011 "Natural Mineral Drinking Water," therapeutic mineral drinking water includes water with total mineralization ranges from 10 to 15 g/l, as well as water with lower mineralization if they contain increased amounts of biologically active substances of iron, boron, arsenic, and other components that have a specific effect in certain diseases. All types of medicinal water should be used only by medical prescription.

Mineral table water includes water with total mineralization from 1 to 10 g/l. *Medicinal-table water* can be used for medicinal purposes (by medical prescription) and can be periodically used as a table drink.

In addition to medicinal mineral drinking and medicinal-table water, natural *table water* is distinguished, which is divided into two types: water with mineralization from 1 to 2 g/l is classified as natural mineral table water, and water with mineralization of less than 1 g/l as natural table water.

In many resorts for drinking mineral water, specialized well rooms are outfitted, and large resorts have large-scale drinking galleries (Fig. 1.14).

Bottled mineral water produced in different regions of the country is very popular. Bottled water usually is artificially saturated with carbon dioxide up to 0.3% (ferruginous—up to 0.4%), which increases its taste and ensures better preservation of the mineral water.

For the production of bottled water, specialized technological lines are used to preserve all the beneficial properties of the water, provide convenience, and increase its life (Fig. 1.15).

To treat mineral water before its bottling, in accordance with GOST, compounds of iron, manganese, sulfur, and arsenic can be separated by filtration, and treated with air

a b

Fig. 1.14 Kislovodsk resort. Entrance to the central drinking gallery (**a**). Mineral water intake (**b**)

Fig. 1.15 A modern technological line for the production of medicinal-table water Shmakovskaya. Primorsky Krai, Skit plant

or oxygen, as well as carbon dioxide saturation. In addition, the use other methods are allowed that do not change the content and ratio of cations (calcium, magnesium, sodium, and potassium), anions (hydrogencarbonates, sulfates, chlorides), and biologically active components in the treated mineral water.

In its organoleptic characteristics, bottled mineral water is a clear, colorless (or with shades from yellowish to greenish) liquid without impurities; natural precipitation of mineral salts is allowed; its taste should reflect the substances contained in water and its level of acidity and alkalinity. The content of toxic elements in bottled mineral water should not exceed established values (Table 1.4).

It should be noted that natural biologically active arsenic contained in medicinal mineral water (Sinegorsk, Chizhepse, etc.) loses its toxic properties. At the same time, since the biological effect of fluorine is still not fully understood, when the

Table 1.4 Acceptable concentration of toxic elements in drinking mineral water

Name of the toxic element	Acceptable concentration of toxic elements, (maximum, in mg/l),		
	Table	Medicinal-table	Medicinal
Barium	1,0	5,0	5,0
Cadmium	0,003	0,003	0,003
Cuprum	1,0	1,0	1,0
Arsenium	0,05	0,05	0,05
Nickel	0,02	0,02	0,02
Nitrates	50,0	50,0	50,0
Nitrites	0,1	0,1	0,1
Mercury	0,001	0,001	0,001
Selenium	0,01	0,05	0,05
Lead	0,01	0,01	0,01
Strontium	7,0	25,0	25,0
Antimony	0,005	0,005	0,005
Chromium	0,05	0,05	0,05
Cyanides	0,07	0,07	0,07

fluoride content in table mineral water is more than 1 mg/l, it is labeled "Contains fluorides"; and for water with a fluoride content of more than 2.0 mg/l — "High fluoride content: not suitable for regular use by children of seven and under".

For external use, mineral water can be used in the form of baths, rinses, compresses, irrigation, etc. The healing effect of water for external use may be due to a local response to the direct effect of water on the skin or mucous membranes, as well as the neuro-reflexive and general humoral response to the physicochemical factors of mineral water.

1.2.2 External (Balneological) Use of Mineral Water

The action of mineral water for balneological use is based on local shifts caused by the influence of physical and chemical factors on the human body. Physical factors include mechanical (the hydrodynamic pressure of water itself and gas bubbles formed in it), thermal (high and low temperatures), and radiation (for radon baths and irrigation) effects. The chemical factor is the gas and ion composition of macro- and microelements, as well as the influence of undissociated molecules (Bogolyubov 2016).

As a result of the effect of these factors on human skin and mucous membranes, a complex adaptive reaction occurs that develops according to the neuro-reflexive and humoral mechanisms due to irritation of thermo-, mechano-, baro-, and chemoreceptors, secretion of biologically active substances, and activation of immune mechanisms. This leads to a weakening of pathological changes, disappearance or reduction of painful and inflammatory phenomena, stimulation of compensatory responses, an increase in the adaptive capacity of the human body, restoration of functions impaired by disease, and improvement of the metabolism and general well-being (Kulakov and Orlansky 1970).

By changing the strength of the temperature stimulus, concentration of natural mineral water and gases dissolved in it, area of exposure (baths, half baths, irrigation, etc.), duration, and number of procedures, you can get different responses from various organs and systems and the whole organism. In addition to dosimetric parameters, the physiological and therapeutic effect of balneotherapy procedures is greatly influenced by the individual characteristics of the patient, his/her reactivity, the initial state of various physiological systems, fitness for temperature effects, age, gender, and profession, as well as the external conditions under which the medicinal procedures are carried out.

1.3 Healing Effect of Mineral Water

For therapeutic purposes, mineral water can be used for various diseases. Currently, Russia has approved ICD-10 (International Statistical Classification of Diseases and Related Health Problems ICD-10 (https://icd.who.int/browse10/2016/en)) standards of spa and resort care for the main nosological forms or corresponding categories of diseases, most of which include use of mineral water for therapeutic purposes. The Ministry of Health of the Russian Federation has approved a list of medical indications and contraindications for a spa treatment (Order of the Ministry of Health of Russia dated 07.06.2018 No. 321n), which outlines general approaches to the use of natural healing factors for various pathologies and regulates the main indications and contraindications. In many diseases, the method of mineral water usage depends on the nature of the pathological process, its severity, stage, and the general condition of the patient, but in most cases the combination of external and internal use of mineral water is most effective. Next, we will consider the possibilities of combined use of drinking and balneological treatment of those categories of diseases where mineral water is in most common use.

To characterize the current health status of the Russian population, statistical material was collected and a database of primary disease incidence per 1000 people for main disease categories according to the ICD-10 between 2003 and 2017 in 85 Russian Federation subjects was created.

Open-source data on primary disease incidence published by the Russian Federal State Statistics Service (https://www.gks.ru) were used as the main source. Further, the regional long-term annual morbidity rates were calculated for each disease category and primary statistical data analysis was conducted.

If data distribution of the variation series was normal, the boundaries of the mapping intervals were calculated by the standard deviation method (for certain infectious and parasitic diseases). If data distribution was not normal (for all other disease categories under study), the quartile method was used. The intervals for morbidity rates for regions were established depending on the variability of values and the boundaries of the statistical standard: average; below/above average; substantially below/substantially above average.

1.3.1 Diseases of the Blood and Blood-Forming Organs and Certain Disorders Involving the Immune Mechanism

In modern literature, the mechanism of the non-specific action of mineral water on the human metabolism and immune system is considered to be one of the main therapeutic factors (Frolkov and Mikhailuk 2014) that affect homeostasis, neuroendocrine regulation, the hematopoietic system, general and tissue immunological reactivity, and the adaptive capacity of the patient body (Bogolyubov 2016). Therewith, the healing effect of mineral water is largely determined by the presence of trace elements in its chemistry.

Ferruginous mineral water. Iron is an integral part of many enzymes and proteins necessary to ensure homeostasis and maintain immunity. The value of ferruginous water is in the high concentration of active (divalent) iron ions. When taken orally, it helps to normalize the blood formation process, improve oxygen transport to tissues, and change the energy supply of metabolic processes in the human body by stimulating the production of red blood cells and increasing hemoglobin levels.

When applied externally, the healing effect of ferruginous mineral water is due to a whole complex of chemical elements contained in it. A bath in ferruginous mineral water has a stimulating effect on blood formation. It is also used for cardiovascular diseases and metabolic disorders, and for irrigation in gynecology.

Carbonated water, including ferruginous, has a stimulating effect on the immune and hematopoietic systems not only through the presence of iron, but also such trace elements as magnesium, copper, and other metals (Komarova and Egorova 1994). These elements are involved in the synthesis of enzymes and proteins that support bone marrow functions in the formation of immune cells (white blood cells, lymphocytes, phagocytes, macrophages, etc.). Carbonated water affects the exchange of nitrogen bases, production of immunoglobulins, and maintenance of the acid–base balance in the human body (Malkov et al. 2015).

Carbonated baths have a normalizing effect on biochemical processes in various body structures in various pathological conditions. From a general biological point of view, response to the effects of carbonated baths can be considered an adaptive reaction of the body, an increase in its resistance to carbon dioxide excess and oxygen deficiency (Oransky 1988; Kneipp 1990; Malyavin 2006).

Arsenic water is used for drinking treatment of anemia, because arsenic is one of the biologically active elements that affect the rate of enzymatic processes of tissue respiration and take part in metabolism (Maloletkina and Ulashchik 1999). For this purpose, arsenic carbonated mineral water with the required concentration (for example, the Chvizhepe deposit at the Sochi resort) is more often used (Vinokurov 2008).

Fluorinated water promotes the active removal of radionuclides and heavy metals from the human body. This is especially desirable for the treatment of hypochromic anemia associated with professional chronic poisoning and toxic liver damage (Vinokurov 2008; Bogolyubov 2016).

Sulfurous (hydrogen sulfide) water. Studies show that low sulfurous water with a sulfide content of not more than 10 mg/l (for example, Pyatigorsk and Sochi water) can act as antioxidants in liver damage, and also have a weak choleretic and diuretic effect (Sheiko and Manshin 2011).

1.3.2 Endocrine, Nutritional, and Metabolic Diseases

Endocrine, nutritional, and metabolic diseases rank second to last in the country based on incidence rates −11.2 per 1000 people in 2003–2017. They include thyroid disorders, diabetes mellitus, nutritional deficiency, and obesity, and other metabolic disorders. This category is characterized by an upward trend: from 9.6 in 2003 to 14.0 in 2017 (Fig. 1.16). An extremely high morbidity rate is observed in Nenets Autonomous Okrug (40.4 per 1000). In Altai Krai (25.3) and Republic of Dagestan (22.7), it is lower but nonetheless notably higher than in other regions. The regions with an incidence rate below average (less than 8.9) are concentrated mainly in European Russia (up to the Udmurt Republic), as well as in regions of the Far East—Khabarovsk and Primorsky Krais, Jewish Autonomous Oblast, and Kamchatka Krai (Map 1.13).

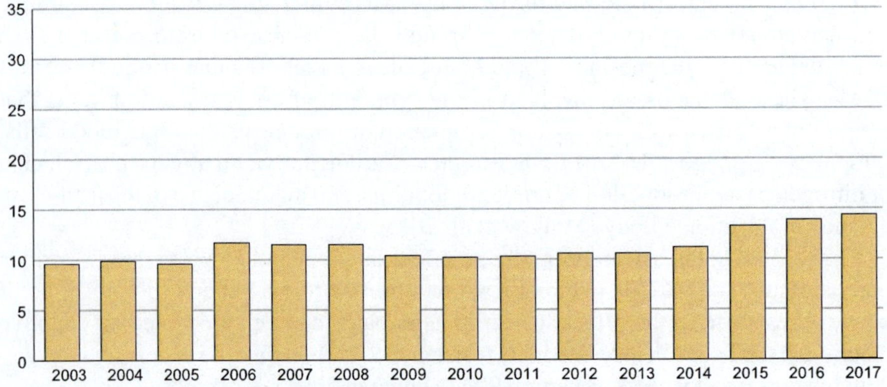

Fig. 1.16 The incidence of the endocrine, nutritional, and metabolic diseases per 1000, 2003–2017

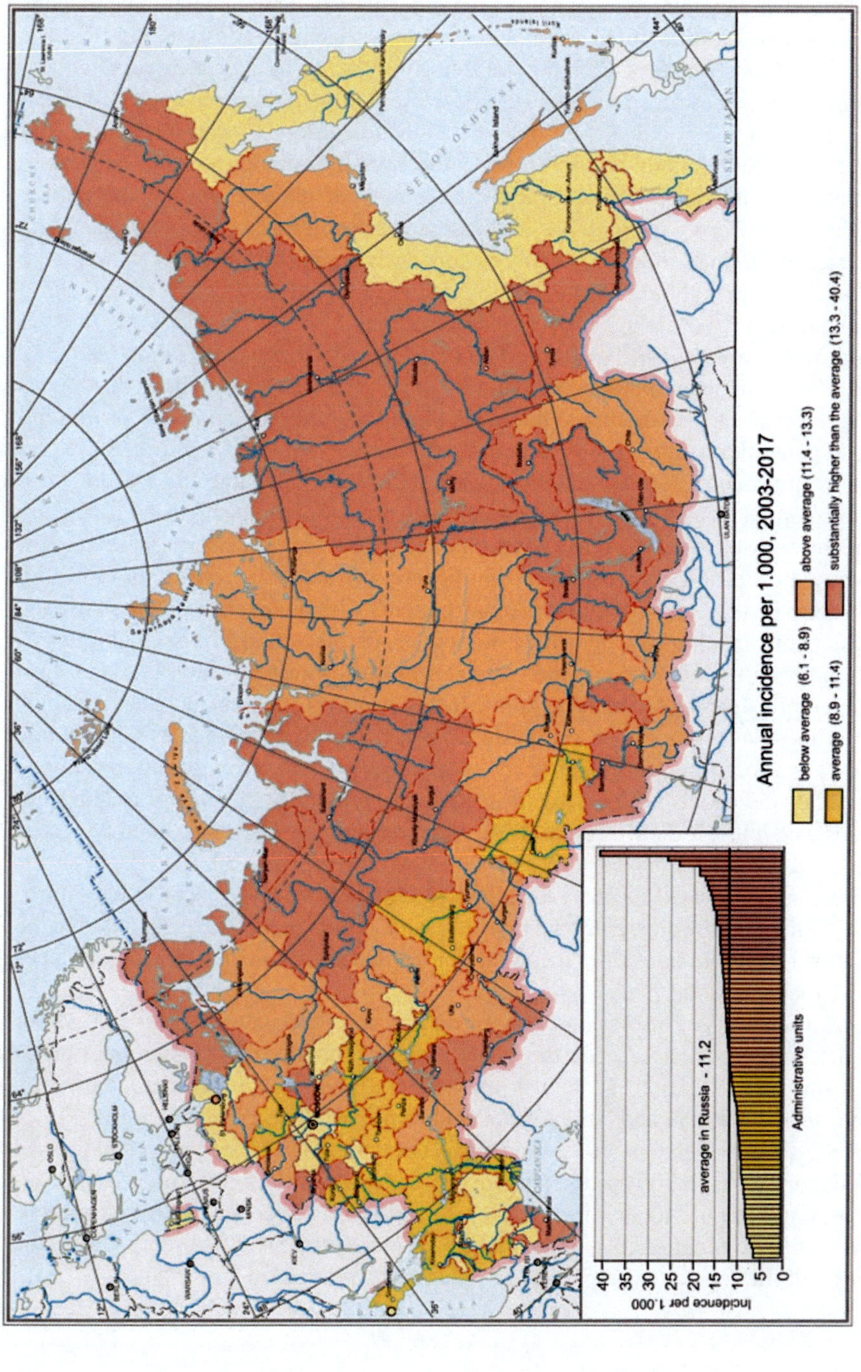

Map 1.13 Endocrine, nutritional, and metabolic diseases, 2003–2017

In recent years, a steady increase in incidence rates is primarily associated with the spread of such disorders as metabolic syndrome, obesity, and diabetes. The reasons for this may be an unfavorable ecological change, sedentary lifestyle, high-calorie nutrition, and other factors affecting the living conditions of a modern person.

Natural mineral water is widely used for the treatment and prevention of such pathologies. In metabolic syndrome, metabolic disorders of carbohydrates, and diabetes mellitus, mineral water improves the course of carbohydrate metabolism, stimulates insulin receptors, and facilitates ingress of glucose into cells (Frolkov and Mikhailuk 2014). In thyroid gland diseases and mineral metabolism disorders associated with endocrine disorders, the minerals found in the water can change the course of the pathological process by filling in an existing deficit or activating adaptive reserves that reduce disease manifestations.

In drinking use of *hydrogencarbonate* and *hydrogencarbonate -sulfate water* in patients with diabetes mellitus, it has a positive effect on metabolism: it reduces hyperglycemia and glucosuria. Therewith the effect of some enzymes promoting glucose penetration into tissues is enhanced, and the lipid spectrum of the blood serum is improved (Vorobev and Parfyonov 1982). At the same time, there is an improvement of tissue carbohydrate metabolism processes, and increasing formation of adenosine triphosphoric acid (ATP), the decay of which releases a large amount of energy. In addition, mineral water significantly increases the sensitivity of insulin receptors to insulin, and therefore, in many patients with diabetes, decreases the need for exogenous insulin and sugar-lowering tablets (Frolkov and Mikhaylyuk 2014).

Magnesium-containing mineral water regulates intestinal motility, and also promotes bile formation and secretion. It has been proven that diabetes mellitus reduces the concentration of magnesium ions in blood by 10–12% (Bogolyubov 2016). Therefore, the additional entry of magnesium ions through internal intake of mineral water makes it possible to sufficiently replenish reserves of magnesium homeostasis and promote activation of metabolic reactions (Sheiko and Manshin 2011).

Iodine water is used for increased thyroid function. Microdoses of iodine, introduced with mineral water, enter the thyroid gland, where the iodine is included in the formation of thyroid hormones triiodothyronine and thyroxine; it ultimately reduces gland hypersecretion.

Carbonated water in various concentrations can stimulate (for 0.6–0.5 g/l) or inhibit (for more than 1.2 g/l) the glucocorticoid function of the adrenal cortex, which must be taken into account in patients with endocrine disorders caused by inhibition or excessive activation of the endocrine glands (Shkolenko 1973).

Nitrogenous water has an analgesic and desensitizing effect, causes changes in hemodynamics, metabolism, the state of the endocrine system, and muscle tone, as well as a number of biochemical blood parameters. In patients with thyrotoxicosis, it stimulates the functionality of the cardiovascular system and favorably affects the function of the regulatory apparatus of the circulatory system.

1.3.3 Diseases of the Nervous System

Incidence of the diseases of the nervous system in 2003–2017 equaled 16 cases per 1000 people. Here, the highest morbidity rates are recorded in the period of 2006–2014 (Fig. 1.17). The morbidity is distributed rather uniformly regionally. There are no territories with outlying values (Map 1.14). Overall, morbidity rates substantially above average (more than 19.8) are observed in some Republics of the North Caucasus—Chechen Republic, Republics of Ingushetia and Dagestan; in Siberia—in Sakha Republic (Yakutia); in the Far East—in Chukotka Autonomous Okrug. The lowest morbidity rates (less than 12.4 per 1000) are recorded in various regions not territorially connected –Tyva Republic, Zabaykalsky Krai, Leningrad Oblast, Jewish Autonomous Oblast, Moscow, Khabarovsk Krai, and Lipetsk Oblast.

The effect of mineral water on the nervous system is largely related to the electrochemical nature of nerve impulse transmission, with the possible influence of trace elements found in mineral water. The main one is sodium. It affects the functioning of the nervous system, because, due to the difference in the concentration of sodium in cells and intercellular fluid, electrical impulses are generated, which is the basis for the functioning of the nervous system.

Chloride sodium water. Baths from this water have a calming and analgesic effect, promote normalization of sleep and mood, encourage a positive commitment to treatment, and restore neurovascular disorders in the spine and limbs (Gurlenya and Bagal 1989). Baths are indicated for diseases of the central (consequences of spinal cord injuries, consequences of poliomyelitis) and peripheral nervous system (plexitis, radiculitis, polyradiculitis).

Nitrogenous water. Dynamics of indicators of nervous and cardiovascular systems, metabolism, etc., observed in patients receiving nitrogen baths indicate that the latter cause a transition from a higher activity level of the body to a relative rest state. This water can be attributed to those weak irritants that have a pronounced ability to enhance inhibitory processes and to relieve excitatory ones (Strelkova 1991).

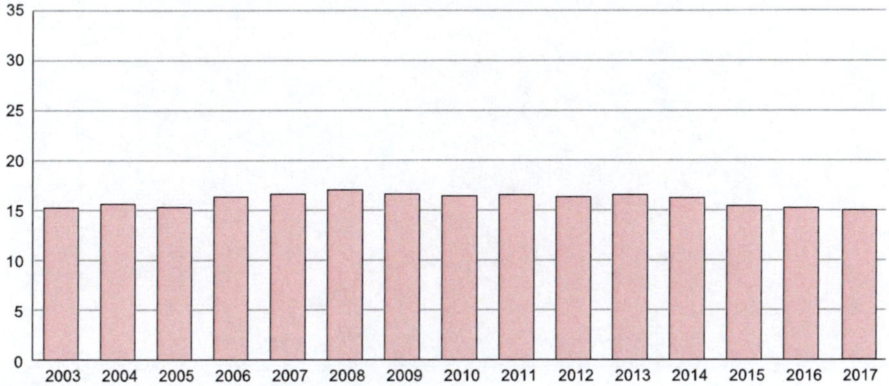

Fig. 1.17 The incidence of the diseases of the nervous system per 1000, 2003–2017

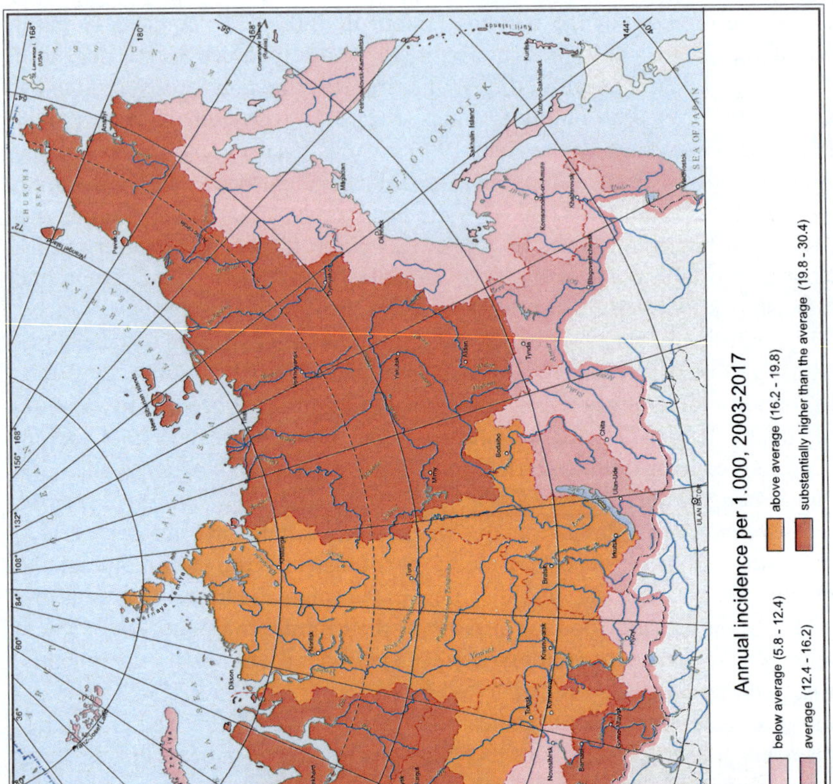

Map 1.14 Diseases of the nervous system, 2003–2017

Fig. 1.18 The use of medicinal iodine-bromine baths for the treatment of vegetative dysfunction of the nervous system in pediatrics. Krasnodar municipal hydropathical establishment

Iodine-bromine baths are indicated for patients with initial manifestations of cerebrovascular insufficiency, as well as diseases of the central and peripheral nervous system such as neuritis, radiculitis, and sciatica, especially with lumbosacral pain syndromes (Fig. 1.18). They promote the restructuring of excitatory-inhibitory processes in the cerebral cortex in the direction of increasing inhibition, causing normalization of muscle tone and a decrease or disappearance of asymmetries in skin temperature and other disorders of the autonomic nervous system. For treatment and prevention of this pathology, iodine-bromine inhalations can be used.

Sulfurous water is used for baths in the treatment of neurological disorders, chronic intoxication with salts of heavy metals (lead, mercury), toxic polyneuritis, encephalitis, etc. Good results were obtained with the use of sulfide baths (50–100 mg/l) in the complex rehabilitation of patients with cerebral atherosclerosis of the elderly (Kozlova et al. 2013). Baths help improve cerebral blood flow, enhance metabolism, increase the vascular tone of the cerebral cortex, and promote the synthesis of neurotransmitters and substances supporting brain functions.

Arsenic baths are recommended for functional neurological disorders. Due to their effect on the metabolism and synthesis of neurotransmitters, processes of excitation and inhibition in the nervous system are normalized, which helps to eliminate the majority of functional neurological disorders. Arsenic ions penetrate intact skin into the internal environment of the body. The plaque formed on the skin during the external use of arsenic baths reduces impulses from skin receptors. It promotes the normalization of excitation and inhibition processes in the central nervous system (Bogolyubov 1986).

Drinking use of hydrogencarbonate sodium-calcium arsenic boron-containing mineral water helps to prevent chronic fatigue syndrome, computer disease, and other pathological responses to electromagnetic fields (Vinokurov 2008).

Radon water is used in the form of baths in the treatment of vertebrogenic radiculitis with severe pain and impaired conduction of impulses along peripheral nerves. Radon concentration in the bath can reach 7.5 kBq/l (200 nCi/l).

In the treatment of patients with traumatic disease of the spinal cord, salt (chloride-sodium, hydrogen sulfide), radon, and gas (nitrogen, oxygen, carbon dioxide) solutions in mineral water are mostly used. The kind of healing effect of mineral water in its external use is determined by the irritating effect of the temperature, and mechanical and chemical stimulating agents contained in water. In the treatment of spinal patients, baths with a temperature of 36–37 °C are used. The so-called indifferent baths contribute to the restoration of temperature sensitivity. It was noted that these baths increase the skin temperature in patients, which indicates that, in a bath with a temperature considered to be indifferent, there is a significant increase in blood circulation in the skin. In the absence of heat exchange between the body and the bath, it leads to equalization of the internal body temperature and the temperature of the skin (Olefirenko 1989; Chernikova 2004).

Hydrogen sulfide water can be used in the form of baths and balneums for spinal patients. Their specificity of action is determined by the body's reaction to hydrogen sulfide and other sulfide ions penetrating the skin. The metabolic rate changes, the endocrine system is stimulated, and vasoactive substances such as acetylcholine and histamine are formed in the skin. A pronounced hyperemia of the skin develops, the amount of blood circulating increases, respiration rhythm and pattern changes, as does oxygen blood saturation, oxidation–reduction reactions accelerate, and protein and water–mineral metabolism increases. A very valuable quality of this water is the intensification of metabolic processes in inflammatory foci. Prescription of hydrogen sulfide baths in the treatment of spinal patients is aimed at improving tissue trophism, stimulating metabolism, accelerating reparative processes, and resorption of inflammatory foci (Karepov 1991).

Internal use of mineral water in the complex therapy of spinal patients makes it possible to actively affect intestinal motility and has a healing effect in urination disorders.

1.3.4 Diseases of the Eye and Adnexa

Diseases of the eye and adnexa include pathologies of the eyelids, eye orbits, lens, retina, as well as glaucoma, diseases of the optic nerve and eye muscles, other visual impairments, and blindness. The dynamics of incidence rate for such pathologies is rather unstable (Fig. 1.19). There, a decline in the number of patients had been observed from 2006 to 2010, in 2011–2012 incidence rates had heightened, and later the number of patients declined again. In respect to territory, varied presentation is observed again; however, based on an analysis of long-term annual average data, territories with unfavorable conditions include, first of all, the regions of Siberia (Map 1.15). Nenets Autonomous Okrug and Altai Republic also stand out sharply in terms of high morbidity rates (more than 60.0 per 1000 people). Below-average morbidity is recorded mainly in the European part of the Russian Federation and in the Far East, as well as in Novosibirsk Oblast and Tyva Republic. The lowest morbidity rate is recorded in Kabardino-Balkar Republic (14.5 per 1000).

The local use of mineral water for eye diseases is justified by its similarity in composition to the tear fluid that naturally moisturizes the eyeball (Fig. 1.20). Irrigation is used for the following diseases and pathological conditions of the eye: conjunctivitis, dry eye syndrome, sty, retinopathy, blepharitis, incipience of cataract, and even glaucoma. Irrigation procedures for the eye, eyeball, and eyelids with different compositions of mineral water are carried out strictly by medical prescription.

There are the following contraindications for this procedure: acute conjunctivitis, ocular mucosa damage, contagious eye diseases, acute infectious diseases, and other general contraindications for a spa treatment (Order No. 321n, 2018).

1.3.5 Diseases of the Circulatory System

Diseases of the circulatory system, including diseases of the heart and blood vessels, hypertension, and other pathologies characterized by impaired arterial, venous, or capillary circulation, are the most common chronic human diseases in Russia. The long-term annual incidence of diseases of the circulatory system equals 27.0 per 1000 people. Here, as opposed to many other disease categories, a stable upward trend is observed (Fig. 1.21). While in 2003 this morbidity rate equaled 20.6, in 2017 it increased to 32.1 per 1000 people. Republic of Ingushetia, Altai Krai, and Chechen Republic (Map 1.16) stand out in terms of high disease incidence (more than 50.0 per 1000). Moreover, the group with a substantially above average incidence level (more than 30.6 per 1000) includes the regions of the Volga territory, the southern Urals, and Western Siberia. The lowest morbidity rates (less than 15.0 per 1000) are recorded in Moscow, Kabardino-Balkar, and Tyva Republics.

The use of mineral water in the treatment of this category of disease has a long history and is associated with its effect both on peripheral (capillary) circulation and on the general (central) mechanisms of regulation of cardiovascular activity and water-salt metabolism in the body.

Carbonic water. Carbonic baths are an active method of functional pathogenetic therapy of circulatory system diseases because they have a stimulating effect on cardiac self-healing and compensation mechanisms, increase myocardial contractile

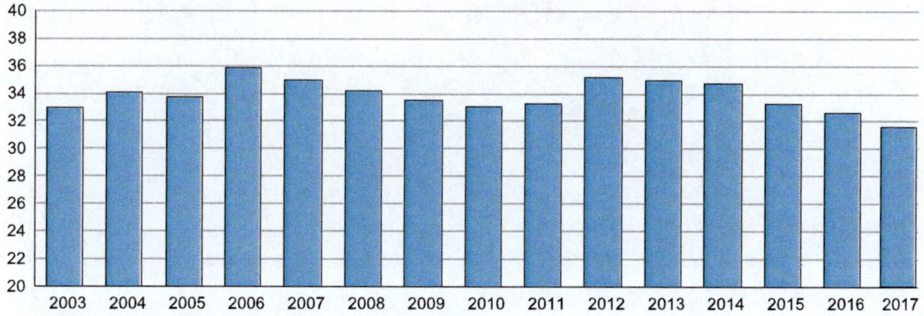

Fig. 1.19 The incidence of the diseases of the eye and adnexa per 1000, 2003–2017

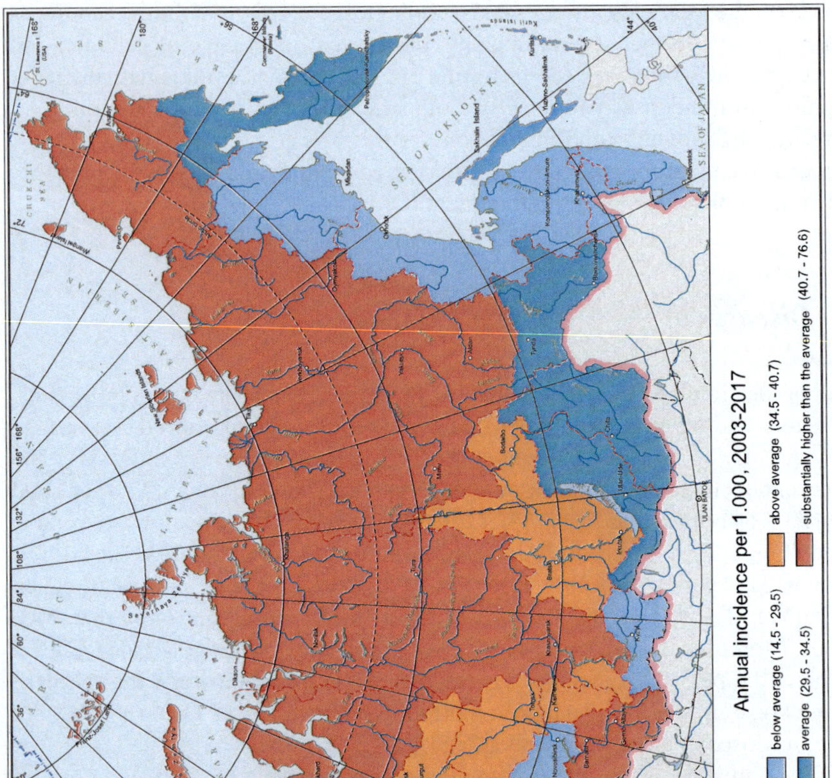

Map 1.15 Diseases of the eye and adnexa, 2003–2017

Fig. 1.20 Eyeball irrigation with mineral water

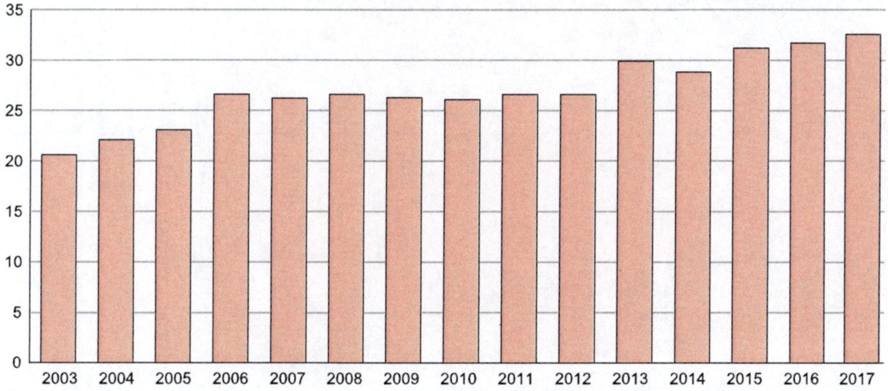

Fig. 1.21 The incidence of the diseases of the circulatory system per 1000, 2003–2017

function, improve cardiac muscle nutrition, and increase the adaptive capacity of the cardiovascular system.

Iodine-bromine water is used for bath treatment of patients with atherosclerotic and myocardial (rheumatic) cardiosclerosis, cerebral atherosclerosis, obliterating endarteritis, and chronic thrombophlebitis (including trophic disorders), valvular heart defects, hypertensive and hypotonic disease, and cardiovascular neurosis.

Sulfurous water is used in the form of a bath. It is prescribed for coronary heart disease, as well as in the short term (1.5–2 months) after myocardial infarction or uncomplicated surgical interventions on coronary vessels (stenting, coronary bypass surgery), if there are no signs of heart failure or circulatory disorders higher than I degree and there is complete healing of postoperative cicatrixes.

Arsenic water is prescribed for hypertension stages I and II, atherosclerosis of the heart and blood vessels, angina pectoris, and chronic coronary heart disease, in the form of a bath. A course of treatment with arsenic water helps to improve central

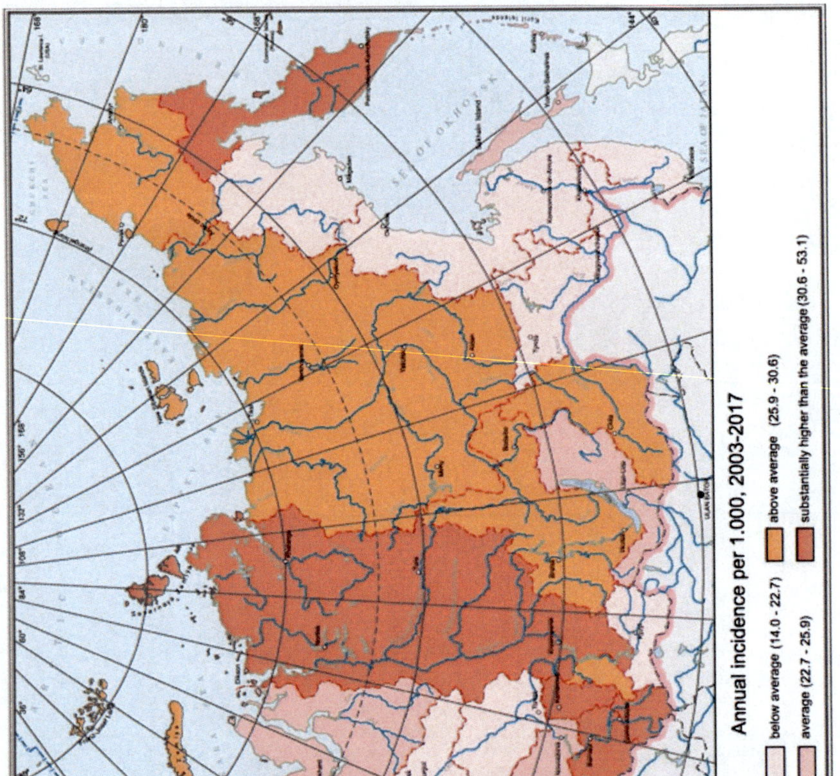

Map 1.16 Diseases of the circulatory system, 2003–2017

and peripheral hemodynamics, improve tissue trophism, accelerate regeneration processes, and increase myocardial contractility (Volkov and Tsikulin 1989).

Radon water: For hypertension, general radon baths are used with a radon concentration of 1.5–3.0–4.5 kBq/l (40–80-120 nCi/l) and water temperature of 35–40 °C, of 10–20 min length, four to five times a week. It is preferable for patients with heart diseases to prescribe radon baths at a lower temperature (35–36 °C) to avoid sharp expansion of capillary vessels and blood redistribution away from the heart. Radon baths are prescribed two to three days in a row or every other day, depending on the general patient condition and state of the circulatory system. In total, 12–15 procedures are prescribed for the course of treatment.

1.3.6 Diseases of the Respiratory System

The greatest number of diseases can be attributed to the diseases of the respiratory system (more than 300 per 1000 people on average in 2003–2017). This category includes such pathologies as influenza and pneumonia, acute respiratory infections of the upper airway, acute and chronic diseases of the lower airway and others. The incidence of the diseases in the respiratory system is characterized by a gradual increase in the number of patients (Fig. 1.22). In 2004–2006, their number did not exceed 300 per 1000 people, whereas in 2017 it reached about 350. Territories with incidence rates substantially above average (more than 371.6) include many northern regions (henceforth administrative units are listed by descending morbidity rate) — Nenets Autonomous Okrug, Chukotka Autonomous Okrug, Yamalo-Nenets Autonomous Okrug, Nenets Autonomous Okrug, Khanty-Mansi Autonomous Okrug – Yugra, Komi, and Sakha (Yakutia) Republics, as well as certain regions of Central Russia — Yaroslavl, Ivanovo, Arkhangelsk, and other Oblasts (Map 1.17). The most favorable situation (less than 285.9) is recorded in the Republics of the North Caucasus — Republics of Adygea and Ingushetia, Karachay-Cherkess, Chechen, and Kabardino-Balkar Republics.

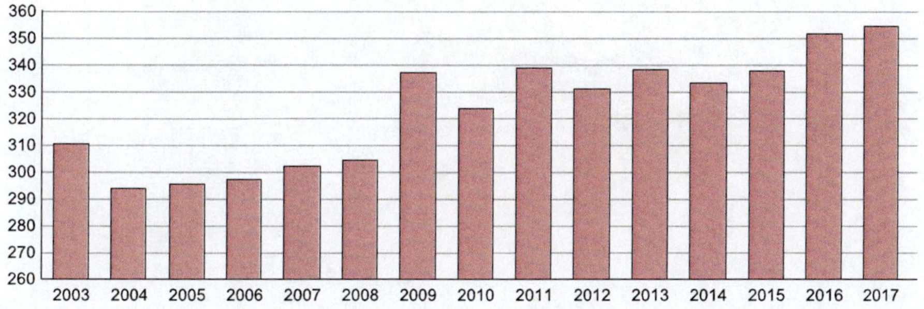

Fig. 1.22 The incidence of the diseases of the respiratory system per 1000, 2003–2017

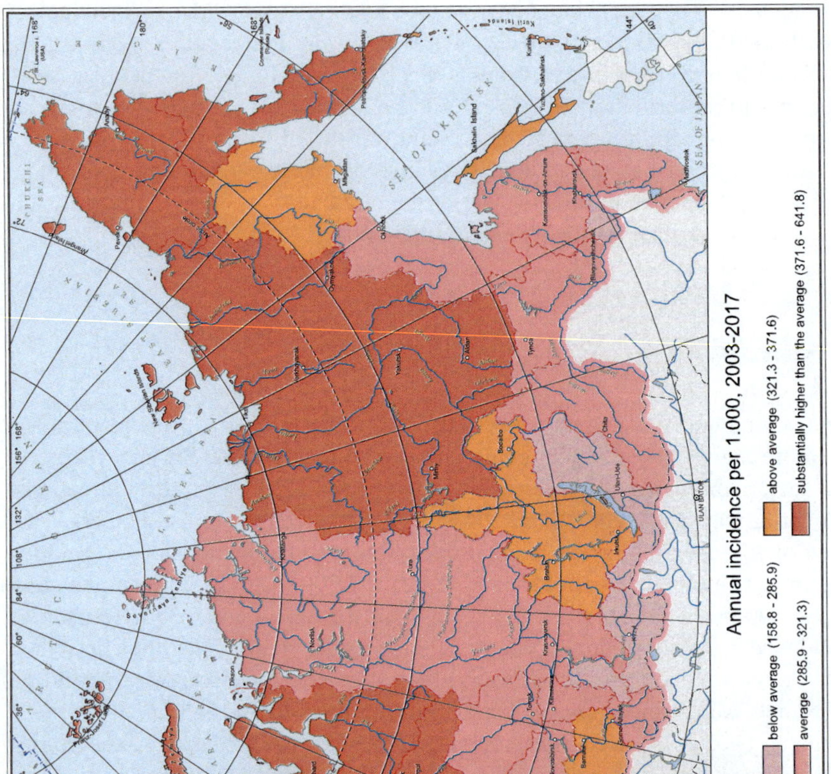

Map 1.17 Diseases of the respiratory system, 2003–2017

Even ancient doctor Galen noticed that people who suffer from coughing feel better during long walks along the sea. He suggested that it is associated with the inhalation of vapors containing sea salt.

Hydrogencarbonate water is prescribed for oral administration in diseases of the upper respiratory tract (chronic laryngitis, tracheitis, bronchitis). The amount of water at one time is 1–1.5 glasses. If the gastrointestinal function is not impaired, the patient can drink water 3–4 times a day 30–45 min before a meal. Water temperature should be in the range of 35–45 °C. Its ingestion helps to increase sputum production, decrease its viscosity, and discharge more freely.

In acute respiratory diseases and upper respiratory tract lesions (chronic, subatrophic and atrophic rhinitis, rhinosinusitis, chronic tonsillitis, ozena, and others), aerosol inhalations are used. For inhalation, low- and medium-mineralized water applicable for drinking treatment is mainly used: hydrogencarbonate sodium (Maykop, Nagutskaya No. 26 and No. 56) and chloride-hydrogencarbonate sodium (Essentuki No. 4); hydrogencarbonate -sulfate calcium and sodium-calcium-magnesium (dolomite Narzan, Narzan, Novoter healing); sulfate and sulfate–chloride with various cationic composition (Borskaya, sulfate Narzan, Krainka, Kashinsk, etc.), as well as calcium-sodium water with various anionic composition (Uglich, Dorokhovo, etc.) (Komarova and Egorova 1994; Grigoryev 2000; Ivanov and Zhuravskaya 2005).

For treatment of respiratory (cold-related) diseases of the upper respiratory tract, irrigation of the nasal mucosa and rinsing of the nasopharynx and throat with mineral or seawater are used (Fig. 1.23). For this purpose, usually, slightly alkaline mineral

Fig. 1.23 Inhalation with hydrogencarbonate—chloride sodium mineral water. Krasnodar Krai, Gelendzhik resort, Lancet Clinic

chloride and hydrogencarbonate water with a medium mineralization level is used (Anapa, Sochi, Staraya Russa, Chapaevskaya, Paratunka). First, in order to prevent an irritating effect, mineral water with a temperature of 36–37 °C is used, and later, water at room temperature can be used for tempering (Kneipp 1990).

In bronchial asthma, non-drug aerosol therapy with various types of mineral water is used, which has not only local, but also a systemic anti-inflammatory effect; it improves the state of the lipid peroxidation system and antioxidant defense, and bronchial patency and adaptive capabilities in patients (Ponomarenko et al. 1998). The greatest clinical effect is achieved by inhalation of hydrogencarbonate, chloride, and low-siliceous water, which can be used in its natural form or diluted to a concentration of 1–2% (Smirnova 2004).

1.3.7 Diseases of the Digestive System

With respect to territory, the incidence of diseases of the digestive system is extremely nonuniform. A slight decrease of diseases of the digestive system was noted only up to 2011, then the value increased again (Fig. 1.24). In most regions of the Urals, Siberia, and the Far East, morbidity rates are higher and notably above average (Map 1.18). The highest rates (more than 58.0 per 1000) are recorded in Omsk and Sakhalin Oblasts, Sakha Republic (Yakutia), Altai Krai, Yamalo-Nenets Autonomous Okrug, Nenets Autonomous Okrug, and Chukotka Autonomous Okrug. Moreover, high incidence rates are a characteristic of Arkhangelsk Oblast, Nenets Autonomous Okrug, and of some regions of the North Caucasus — Republics of Dagestan and Ingushetia. Below-average morbidity is observed mainly in European Russia, with minimal values (less than 19.0) in Kirov Oblast, the city of Moscow, Republic of North Ossetia (Alania), Lipetsk Oblast, and Stavropol Krai.

For historical reasons, the use of mineral water for diseases of the gastrointestinal system is the most studied. A large number of scientific works and experimental studies are devoted to this issue (Shkolenko 1973; Illarionov 2004). In certain diseases, the method of water ingestion may vary significantly.

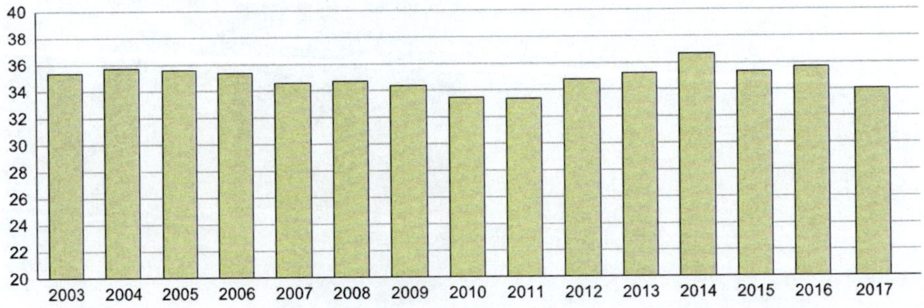

Fig. 1.24 The incidence of the diseases of the digestive system per 1000, 2003–2017

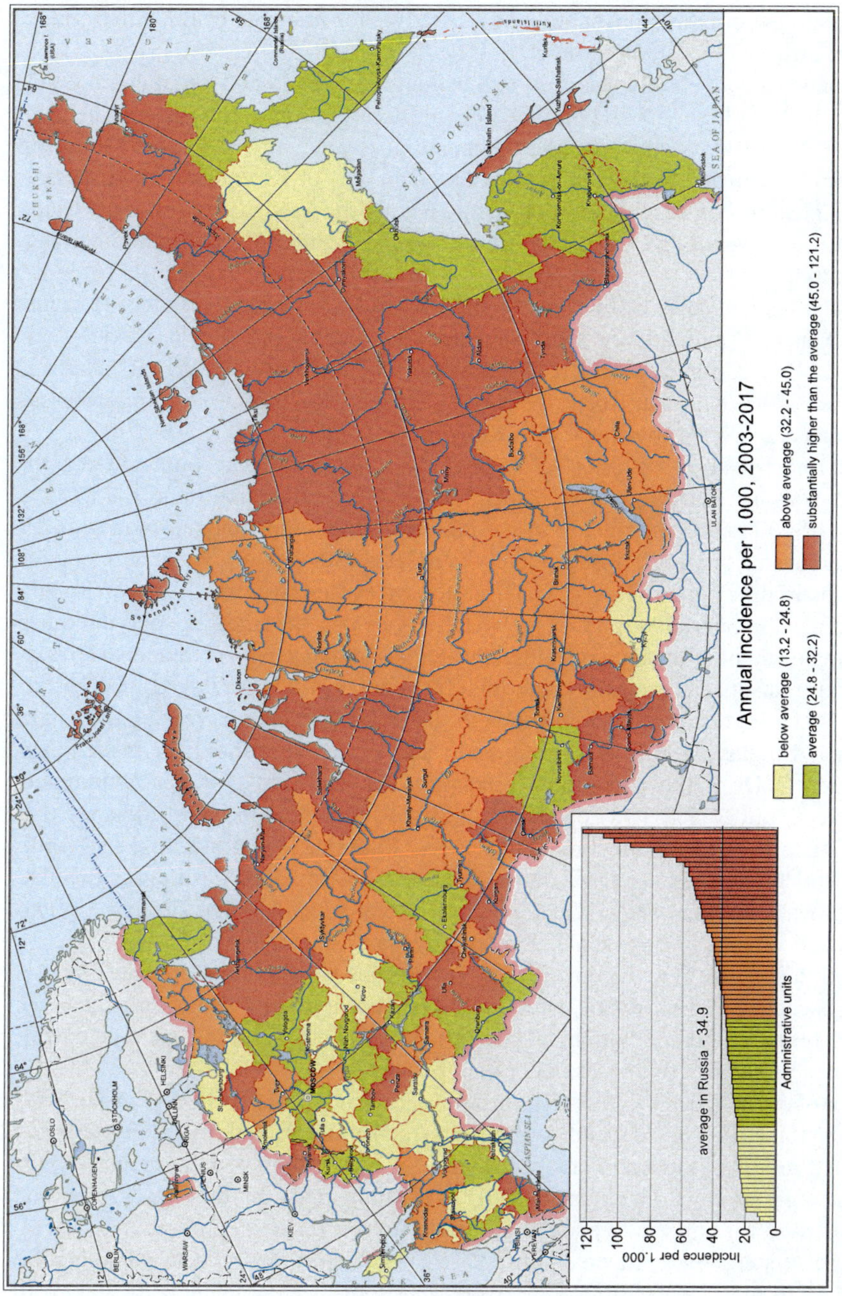

Map 1.18 Diseases of the digestive system, 2003–2017

For drinking treatment of gastrointestinal diseases, low-mineralized (2–5 g/l) and medium-mineralized (5–15 g/l) water is used. Highly mineralized water (15–35 g/l) is used for concomitant constipation as a laxative, for drainage of the biliary tract, and for nutritional obesity treatment.

For patients with chronic gastritis with increased secretory activity, mineral water is prescribed 3–4 times a day, 1–1.5 h before meals. Water ingestion is recommended to start with 0.5 glass, and increase the single dose gradually to 1 glass (180–200 ml). Drink water quickly, in large sips. The main part of the mineral water passes into the intestine before meals, and therefore mainly the duodenal effect of the water is manifested, associated with inhibition of gastric secretion. The optimum water temperature is 38–45 °C. When heated, excess carbon dioxide is released from the water, and heated water irritates nerve endings of the gastric mucosa less (Kipiani 1974). For such patients, the following mineral water is used: Essentuki No. 4, Annensk, Buy, Goryachiy Kluch, Gusev, Darasun, Kashinsk, Kislovodsk Narzan, Kozhanovskaya, Krainka, Kuka, Lazarevsk, Lipetsk, Novoizhevsk, Staraya Russa, Slavyanovskaya, Smirnovskaya, Ufa, Khilovo, and others.

In chronic gastritis with normal acidity, the same water is recommended as in gastritis with increased secretion. Mineral water is prescribed the same 3–4 times a day, but 45–60 min before meals. Water ingestion is recommended to start with 0.5 glass, increasing the single dose gradually to 1 glass (180–200 ml). Water temperature should be in the range of 28–35 °C. Drink the water slowly, in small sips. With this method of administration, part of the water passes into the intestine, and the other part remains in the stomach, i.e., mineral water has both gastric and duodenal effects.

In chronic subacid gastritis, drink mineral water 3–4 times a day 20 min before meals. Water ingestion is recommended to start with 0.5 glass, increasing the single dose to 1–1.5 glass (180–250 ml). Mineral water temperature should be in the range of 18–25 °C. Drink the water slowly, in small sips. This intake of water stimulates stomach secretory activity. For chronic gastritis with secretory insufficiency, the following mineral water is recommended to use: Essentuki No. 4 and No. 17, Mashuk No. 1 and No. 19, Azov, Daut Narzan, Izhevsk, Karmadon, Lazarevsk, Lipetsk, Mikhailovskaya, Nagutskaya, Obukhovskaya, Solonikha, Sloboda, Tazalyk, Uglich, Khilovo, Feodosia, Khadyzhensk, Shmakovskaya, Iasi, etc.

Drinking treatment with mineral water is prescribed mainly for non-acute diseases in the absence of a hemorrhagic tendency and persistent pyloric stenosis. For diseases of the stomach after operation, in the absence of complications, treatment with mineral water begins two weeks after the operation.

Treatment of chronic enteritis, chronic colitis, and intestinal dyskinesia with drinking mineral water depends on the intestinal motility pattern. With increased intestinal motility (intestinal cramps, diarrhea), water with a small amount of gas, low and medium mineralization (2–8 g/l), is recommended. In these cases, the same water is used for chronic hyperacid gastritis (Shkolenko 1973; Bogolyubov 2016). Drink the water slowly, in small sips, 35–60 min before meals. Water temperature should be in the range of 40–45 °C. Water ingestion is recommended to start with 0.5 glass, increasing the single dose gradually to 180–220 ml. Mineral water helps to thin mucus and remove it from the intestinal mucosa. Calcium ions increase the permeability of the cell membranes and enhance the anti-inflammatory effect. Hot degassed water relaxes the intestinal smooth muscles, thereby eliminating spastic phenomena. In case of spastic constipation, warm water helps to eliminate spastic phenomena and pain, and promotes regular bowel evacuation.

For chronic enteritis, chronic colitis manifested as dyskinesia with reduced motion function (atony), mineral water with medium and high mineralization (6–20 g/l) is used. It is recommended to drink this water 3–4 times a day 0.5 h before meals. Water temperature should be in the range of 18–24 °C. Drink water quickly, in large sips (Bogolyubov 2016). The same water as for chronic subacid gastritis will be used. In addition, sulfate magnesium-sodium and sulfate–chloride sodium-calcium water have a good laxative effect (sulfate Narzan, Aksai, Vyatskie Polyany, Dorokhovo, Zvenigorod, Kashinsky, Kudepsta, Lagonaki, Pacha-Yavlakh, Surok, Solonikha, Cheboksary). These types of water help to slow fluid absorption in the intestine, and high mineralization water with this method of administration helps to increase intestinal tone and peristaltic motions (Sahakyan 1975).

For chronic diseases of the liver and biliary tract, various mineral water is used (Essentuki No. 4 and No. 17, Amur, Batalinskaya, Darasun, Karmadon, Kashinsk, Krainka, Lazarevsk, Malka, Nagutskaya, Psekups, Sernovodskaya, Slavyanovskaya, Smirnovskaya, Uglich, Kuka, Ufa, Feodosia, Khilovo, Shmakovskaya, etc.). Water intake is prescribed 3–4 times a day, water temperature should be 35–45 °C. The time and method of its administration depend on the state of the stomach secretory function.

Mineral water has a normalizing effect on the liver function and bile secretion, increases bile production by liver cells, improves bile composition and carbohydrate and protein metabolism in the liver, and enhances its protective function. In gallbladder diseases, mineral water entering the duodenum causes a bladder reflex, which is expressed as gallbladder contraction and increasing in the peristalsis of the biliary tract (Shkolenko 1973; Shautsukova 1976). It helps to empty the gallbladder and improve the flow of bile into the intestines. Thereby, favorable conditions are created for eliminating stagnation and inflammatory phenomena in the gallbladder and biliary tract. Mineral water prevents the precipitation of cholesterol crystals, which can accumulate in the gallbladder and bile ducts and cause gallstone formation (Malkov et al. 2015).

Along with drinking treatment of gastrointestinal diseases, other methods of internal use of mineral water are also used, which include: gum irrigation, gastric lavage, duodenal intubation, duodenal drainage, transduodenal lavage, intubation, and microclysters. Gastric lavage procedures are undertaken from one to three times a week, mainly with low mineralized water at a temperature of 40–42 °C; the amount of water per procedure is from two to five liters.

In case of biliary tract dysfunction with peristalsis insufficiency (hypotonic type), for peristalsis enhancement, as well as with hypertonic-type dyskinesia (excessive peristalsis), for reducing of spastic phenomena and bile dilution, bile duct drainage is performed without a probe (tubage). This procedure improves bile flow into the intestines. For this purpose, the patient drinks 0.4–0.5 l of mineral water at a temperature of 42–45 °C in small sips while lying on his/her right side, or in a reclining posture on his/her back. After this, it is recommended that the patient lies on the right side for 45–60 min; a warm heating pad may be placed on the right hypochondrium (Manshina 2007). Treatment should be provided by medical prescription under mandatory medical supervision. Intubation may be carried out routinely for several months.

In chronic pancreatitis, the same mineral water is used as in diseases of the liver and biliary tract. Water type, its temperature, and time of intake depend on the state of the stomach secretory function. Mineral water has a positive effect on the enzymatic processes of the body, and improves pancreatic activity. In case of insufficient pancreatic function, mineral water is prescribed as in chronic subacid gastritis (Shkolenko 1973). With excessive exocrine activity, the same water is used as in chronic hyperacid gastritis.

For gastrointestinal diseases, medicinal mineral baths are also used. Iodine-bromine baths have a harmonizing effect on the autonomic nervous system, as well as anti-inflammatory and sedative effects. Hydrogen-sulfide or sulfide baths also have a pronounced sedative effect. Strong sulfide water is contraindicated in gastric ulcer and peptic ulcer disease, and chronic diseases of the liver and biliary tract with a tendency toward exacerbation. Balneotherapy with carbon dioxide-arsenic water is effective for chronic cholecystitis, biliary dyskinesia, chronic gastritis, and enterocolitis. Radon baths are prescribed for biliary dyskinesia. After a course of treatment with radon baths, gallbladder contractility improves: biliary tract hypotension decreases in frequency and severity (Sheiko and Manshin 2011).

1.3.8 Diseases of the Skin and Subcutaneous Tissue

Diseases of the skin and subcutaneous tissue rank third in Russia in terms of disease incidence (47.4 per 1000 people in 2003–2017). They include various skin infections, dermatitis and eczema, hives, and skin diseases associated with irradiation (ultraviolet and other). The morbidity dynamics (Fig. 1.25) reflects a downward trend; the number of new cases, starting from 2006 (50.0 per 1000), declines progressively

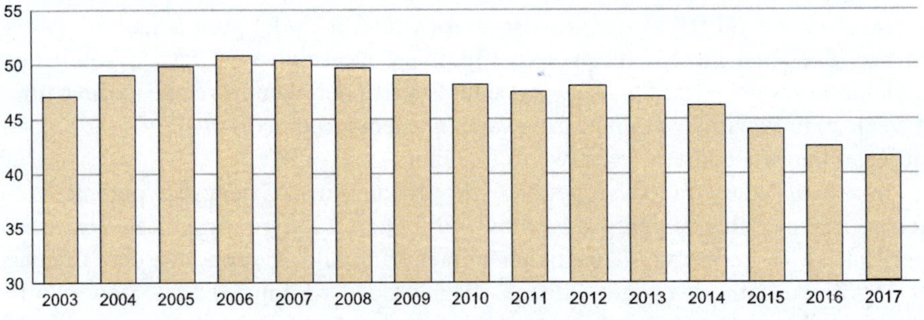

Fig. 1.25 The incidence of the diseases of the skin and subcutaneous tissue per 1000, 2003–2017

(41.0 per 1000 in 2017). There are no clearly marked territorial confines for this disease category (Map 1.19). Five regions with the highest incidence rates (more than 65.0 per 1000 people) include Nenets Autonomous Okrug, Komi Republic, Republic of Karelia, Chuvash Republic, and Vladimir Oblast; the lowest (less than 32.0 per 1000) are in Kabardino-Balkar Republic, Stavropol Krai, Voronezh Oblast, Republic of Buryatia, and Chechen Republic.

It is widely known that skin diseases are catalyzing factors for many disorders of the gastrointestinal tract, kidneys, liver, and immune and lymphatic systems. A reaction in the form of color or structural changes of the skin tissue clearly demonstrates manifestations of internal body problems. The outer cover has a protective function against physical and chemical influences, and removes metabolic products, slags, toxins, and harmful substances (Dovzhansky and Orzheshkovsky 1986).

The treatment of skin diseases using mineral water has been known for a long time. Considering the complex, multicomponent nature of skin disease pathogenesis, the complex use of mineral water is advisable in their treatment, both for external and drinking use. Indications for the use of mineral water are psoriasis, neurodermatitis, eczema, rosacea, acne, dermatitis, papulosquamous disorders, hives, erythema, hyperkeratosis, and other diseases of the skin and subcutaneous tissue.

Highly mineralized natural mineral water of open reservoirs (Tambukan Lake, Saki Lake, Elton Lake, etc.) and natural thermal springs (Paratunka, Malki, Melchhi, Matsesta, etc.) is widely used as an external treatment. The greatest effect is observed with the combined use of mineral water and therapeutic mud, due to which inflammatory manifestations and itching are reduced, and damaged skin is gradually regenerated (Sosin and Buyavykh 2001).

1.3.9 Diseases of the Musculoskeletal System and Connective Tissue

Diseases of the musculoskeletal system and connective tissue rank eighth in the morbidity level. Overall, there is a downward trend in their incidence rates (Fig. 1.26) — from the highest value in 2006 (35.4 per 1000 people) to the lowest in 2017 (29.5). Nenets Autonomous Okrug stands out in terms of morbidity rate (79.1), where it is almost twice as high as in other subjects of the Russian Federation (Map 1.20). Also, this incidence rate is significantly above average (more than 39.9 per 1000) in such regions as Altai Krai, Chukotka Autonomous Okrug, Irkutsk Oblast, Republic of Karelia, Oryol Oblast, and some others. It is below average (less than 27.1 per 1000) in Voronezh Oblast, Chechen Republic, Stavropol Krai, Tyva Republic, and others, as well as in Kabardino-Balkar Republic, which is traditionally characterized by minimal incidence rates for this disease category.

Diseases of the locomotive apparatus and connective tissue are one of the main reasons for referring patients to spa resort treatment (Bogolyubov 1986). In inflammatory and degenerative diseases of the joints and spine, chloride sodium water is used.

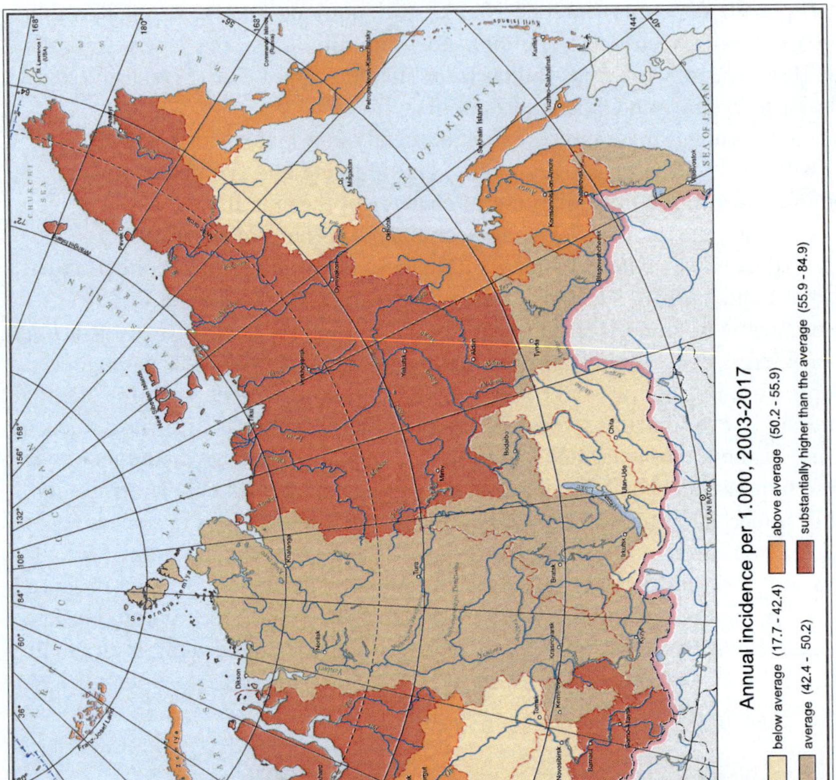

Map 1.19 Diseases of the skin and subcutaneous tissue, 2003–2017

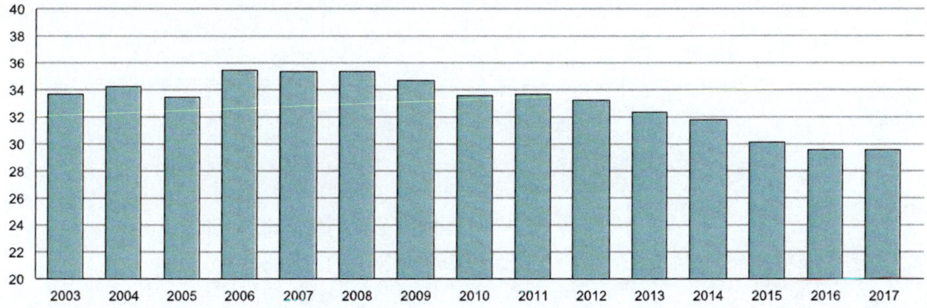

Fig. 1.26 The incidence of the diseases of the musculoskeletal system and connective tissue per 1000, 2003–2017

Salt baths in musculoskeletal diseases help to relax muscles, reduce swelling, improve microcirculation and tissue trophism, eliminate hypercoagulation of blood, normalize immunological processes, relieve pain in the joints and spine, and have an overall anti-inflammatory effect. These effects are enhanced at a water temperature of 30–35 °C (Shautsukova 1976). Normalizing blood circulation and metabolism in the skin and soft tissues, saltwater has a restorative and tonic effect, and it balances excitation and inhibition processes in the central nervous system, which helps to improve mood and increase working efficiency. Salt baths have desensitizing, anti-inflammatory, trophic and analgesic effects (Kozlova et al. 2013; Bogolyubov 2016).

A course of treatment includes 5 to 20 procedures, which can be carried out sequentially in several stages: first, in the form of balneotherapy (salt baths), then in the form of hydrotherapy (underwater massage and exercise in salt or seawater), and culminate by thalassotherapy (bathing and therapeutic swimming in the sea). For degenerative dystrophic diseases of the spine (dorsopathies), underwater traction can be used at the initial stage by medical prescription (Starikov et al. 2013).

Iodine-bromine water, most often with a sodium chloride composition, is indicated for a combination of arthrosis, arthritis, dorsopathy, and other diseases of the musculoskeletal system with circulatory diseases, metabolic disorders, and endocrine dysfunctions. In consequence of the sedative effect of bromine, these baths are useful for patients suffering from emotional instability, insomnia, and anxiety. Iodine-bromine baths are indicated for patients with a mild form of rheumatoid arthritis, in the acute stage of gouty arthritis, or ankylosing spondylitis with minimal activity. A resort treatment course includes 10 to 12 procedures, daily or on alternate days. Iodine-bromine baths have desensitizing, immuno-correcting, anti-inflammatory, trophic, and analgesic effects (Bogolyubov 1986).

For musculoskeletal diseases, siliceous and nitrogen-siliceous mineral water is also used for balneotherapeutic procedures (Kuldur, Goryachinsk, Paratunka, Annensk, Tumnin, Byssinsky, Malkinsky, etc.). The metasilicic acid contained in these water types has adsorption, anti-inflammatory, and analgesic properties that contribute to the restoration of cartilage and bone tissue, and nitrogen dissolved in the water has a desensitizing effect, improves microcirculation, and helps to reduce

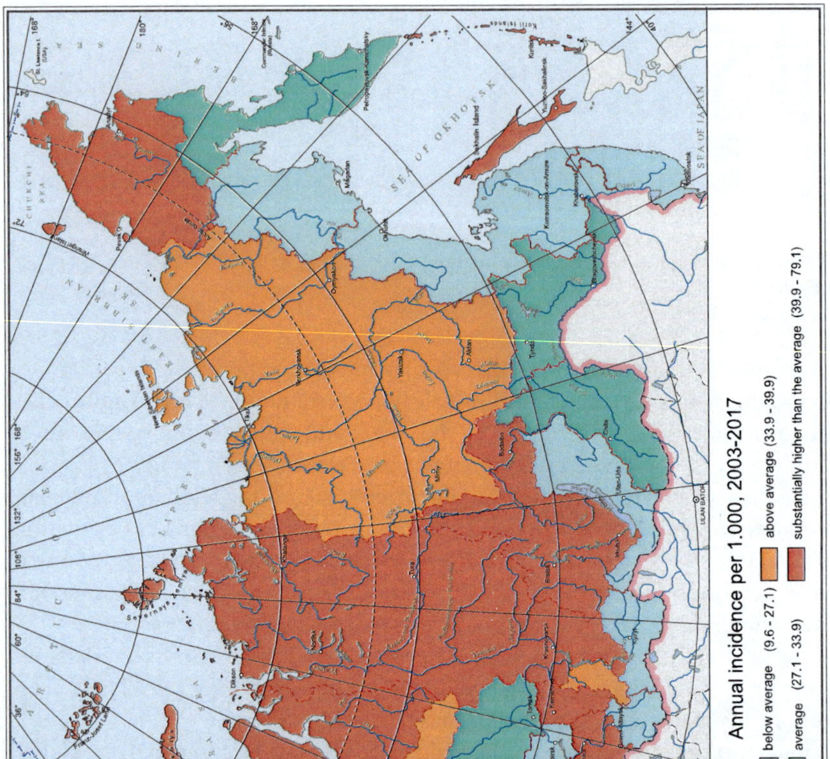

Map 1.20 Diseases of the musculoskeletal system and connective tissue, 2003–2017

tone, restore joint mobility and normalize metabolism in muscle and connective tissue. The resort treatment course includes 10 to 12 procedures, daily or on alternate days. For joints of the upper and lower limbs, local (chamber) baths can be used.

Hydrogen sulfide (sulfide) water is indicated (in the form of local and general baths, irrigation, and showers) for rheumatoid arthritis in the inactive stage or with minimal activity, as well as for degenerative and metabolic diseases of the spine and joints. Hydrogen sulfide drastically changes peripheral blood circulation, which is manifested in severe skin hyperemia at the site of local exposure. It improves blood circulation and metabolic processes in the skin and underlying tissues, including muscles, joint capsules, and cartilage. Hydrogen sulfide enhances the thermal effect of medicinal baths through increasing oxygen absorption, which, in combination with activation of the blood supply to tissues and increasing of enzymatic activity, activates tissue repair and peripheral nerve regeneration, accelerates the healing of bone fractures, and streamlines the structure of collagen fibers in scars, increasing their distensibility and elasticity. Hydrogen sulfide baths and balneotherapy have the following effects: desensitizing, immuno-correcting, anti-inflammatory, trophic, analgesic, restoring impaired joint function, and preventing disease progression. The duration of the analgesic effect in patients with osteoarthritis, rheumatoid arthritis, fibromyalgia, ankylosing spondylitis, dorsopathies, and back pain is usually three to six months. A resort treatment course includes 8–12 procedures, which are carried out every other day, or two days in a row, with a break on the third day. Local hydrogen sulfide baths of 2 and 4 chambers (for arms and legs) are prescribed to patients with a decrease in functional reserves of the cardiovascular system.

Radon water is effective for inflammatory and degenerative diseases of the joints and spine, gout, as well as in autoimmune diseases of the musculoskeletal system (rheumatoid arthritis, ankylosing spondylitis). Radon water is used in the form of radon baths, irrigation, and in therapeutic pools. Radonotherapy has the following effects: desensitizing, immuno-correcting, anti-inflammatory, trophic, analgesic, restoring impaired joint function, and preventing disease progression (Bogolyubov 2016). The analgesic effect after resort spa treatment using radonotherapy can increase for several months and lasts up to a year. A course of treatment includes 8–12 radon baths and other treatment procedures with radon water, which are carried out daily or on alternate days. Repeated courses of radon baths are recommended no earlier than after six months (Veselovsky et al. 1982).

1.3.10 Diseases of the Genitourinary System

The incidence rate of diseases of the genitourinary system is similar to the long-term annual average incidence rate of diseases of the skin and subcutaneous tissue (47.4 per 1000 people). This category includes such pathologies as renal impairment, urinary stone disease, diseases of the male reproductive organs, diseases of the mammary glands, diseases of the female pelvic organs, and others. In 2003–2017 (Fig. 1.27), a

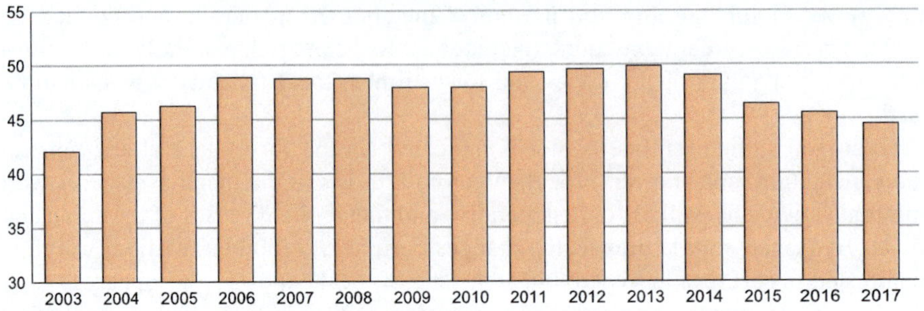

Fig. 1.27 The incidence of the diseases of the genitourinary system per 1000, 2003–2017

waveform incidence dynamic with two elevations in 2006 and 2013 (48.9 and 49.8, correspondingly) was recorded. There is a rather prominent group of the northern regions with morbidity levels notably above average (more than 56.5 per 1000) — that is, Nenets Autonomous Okrug, Yamalo-Nenets Autonomous Okrug, Republic of Karelia, Komi Republic, Arkhangelsk Oblast, and Khanty-Mansi Autonomous Okrug – Yugra (Map 1.21). Along with the Republics of the North Caucasus, the morbidity rate is significantly below average (less than 39.0 per 1000) in Pskov, Moscow, and Kursk Oblasts.

Mineral water is widely used in diseases of the kidneys and urinary tract (except for tuberculous ones). In case of salt metabolism disorder, the purpose of using of drinking mineral water depends on the reaction or type of disorder that caused the formation of urinary concrements. It is commonly known that urinary concrements dissolve in an alkaline environment (urates, oxalates) or in acidic (phosphates). Treatment with mineral water is carried out only with small concrements that aren't a serious obstacle to urine outflow and can freely pass through the urinary tract. Drinking treatment is not recommended in the presence of large concrements. Mineral water is ingested six times a day before meals and 2–3 h after meals. One ingestion is one or two glasses (180–360 ml) of water with a temperature of 24–30 °C (Bogolyubov 2016). In urinary tract diseases, the method of drinking treatment depends on the state of the stomach secretory function.

Mineral water increases urination, promotes dissolution and excretion of mucus from the urinary tract; normalizes urine composition; reduces inflammation, intoxication, and pain; promotes the removal of urocheras and sometimes the elimination of urinary concrements. This applies only to concrements formed from uric acid. There is only limited evidence that mineral water can completely dissolve concrements in the urinary tract. For the treatment of chronic pyelonephritis and urolithiasis, mineral water is used of such types as: Essentuki No. 4 and No. 17, Amur, Arzhi, Borskaya, Darasun, Dolomite Narzan, Kashin Resort, Krainka, Kozhanovskaya, Nagutskaya No. 4, Narzan, Slavyanovskaya, Smirnovskaya, Ufa, Khilovo, and others.

For the urolithiasis treatment, acidic low-mineralized or hydrogencarbonate calcium or magnesium water is most popular. Alkaline hydrogencarbonate calcium or magnesium water is usually used when urinary alkalization is needed. It should

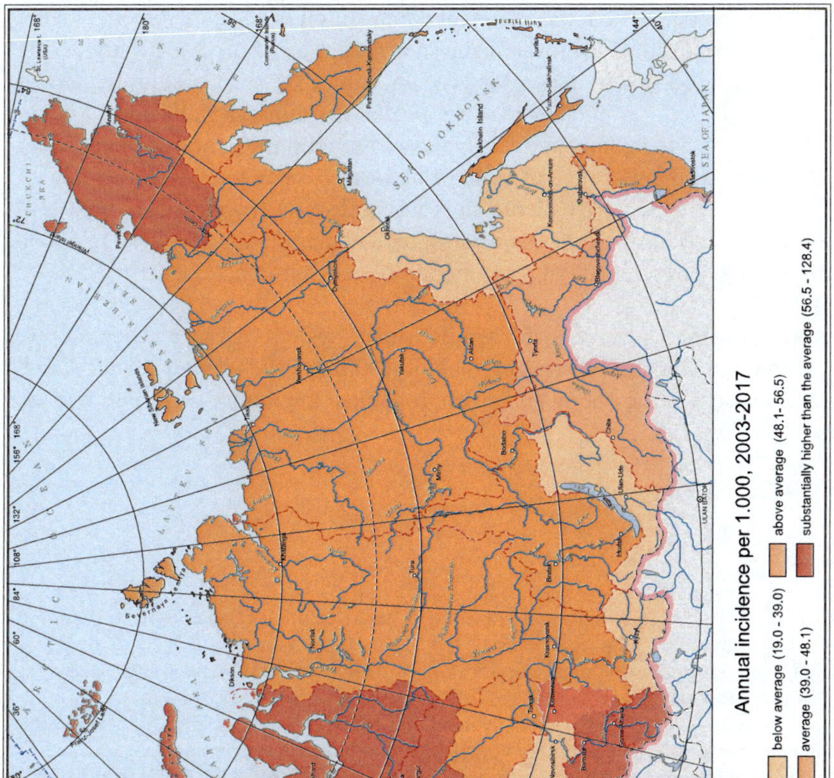

Map 1.21 Diseases of the genitourinary system, 2003–2017

be remembered that, when drinking mineral water, urine pH changes faster than when prescribing a diet (Bogolyubov 2016). With urolithiasis, the urine pH should be adjusted in accordance with the chemical composition of the concrements. In the presence of urate or cystine, urine alkalization is necessary; with oxalate stones, pH should not be changed, and in the presence of phosphate stones, acidic water should be recommended. Sulfates and magnesium present in mineral water have a laxative effect, which is undesirable since it increases the proportion of extrarenal water excretion. However, the presence of magnesium ions is beneficial for patients with a high oxalate level. Some trace elements in mineral water (copper, iron, tungsten) promote the dissolution of oxalate and phosphate concrements. The available data on the dissolution effect of magnetized mineral water on concrements need to be verified.

Iodine-bromine baths are used for genitourinary system diseases. They help to improve blood supply to internal organs, including kidneys and adrenal glands, which lead to an increase in urine output and an improvement in glomerular filtration in the kidneys (Mavritsky 1985).

References

Aleksandrov VA (1932) The basics of balneology. Multi-volume manual. Vol. I. Moscow: Meditsinskaya Literatura, 500p (in Russian)

Alekin OA (1948) Revisiting chemical classification of natural waters. Trudy NIU GUGMS Ser. 4(32):25–39 (in Russian)

Badalov NG, Konchugova TV, Persiyakova-Dubova MLA, AL, Martynova EYu, (2016) The role of non-pharmacological methods in the complex of measures for prevention and treatment of osteoporosis. Sovremennaya Revmatologiya (modern Rheumatology) 3:62–66 (in Russian)

Bogolyubov VM (ed.) (1986) Handbook of health-resort treatment selection. Moscow: Meditsina, 525p (in Russian)

Bulyenkov NA (2005) The role of modular design in study of systemic self-organization processes. Biophysics 50(5):620–644 (in Russian)

Bogolyubov VM (ed.) (2016) Physiotherapy and balneology. Book 3. Moscow: Binom 312p (in Russian)

Chernikova LA (2004) Neurorehabilitation: problems and solutions. Fizioterapiya. Bal'neologiya i Reabilitatsiya (russian Journal of Physiotherapy, Balneology and Rehabilitation) 3:3–10 (in Russian)

Dovzhansky SI, Orzheshkovsky VV (1986) Physiotherapy of skin diseases. Saratov: Universitet, 200p (in Russian)

Frolkov VK, Mikhaylyuk OV (2014) Natural and physical factors in metabolism correction in patients with metabolic syndrome. Fizioterapiya. Bal'neologiya i Reabilitatsiya (russian Journal of Physiotherapy, Balneology and Rehabilitation) 4:11–14 (in Russian)

Gall LN (2015) Matter and life. St. Petersburg: Amphora, 319p (in Russian)

Gerasimenko MYu, Burakov AYu, Badalov NG (2016) Balneological opinion in regard of natural mineral drinking medicinal table water from wells No. 1, 2, 3 and 4. Sanatorium Martial Waters LLC in Kondopozhsky District of the Republic of Karelia dated 19.05.2016 No. CC-34. Approved by the director of FSBI RNC MRiK, 4p (in Russian)

Grigoryev SP (2000) Aeroinotherapy in treatment of patients with bronchopulmonary pathology. Rossiiskii Meditsinskii Zhurnal (medical Journal of the Russian Federation) 6:44–47 (in Russian)

Gurlenya AM, Bagal GE (1989) Physiotherapy and balneology of nervous diseases. Minsk: Visheyshaya Shkola, 398p (in Russian)
Ivanov VV (ed) (1968) Map of mineral healing waters of the USSR
Ivanov VV, Nevraev GA (1964) Classification of underground mineral waters of the USSR. Moscow: Nedra Publishers, 168p (in Russian)
Ivanova VV (ed.) (1974) Mineral waters of the USSR. Explanatory note to the map of mineral waters of the USSR, scale 1: 4 000 000. Moscow: CNII kurortologii i fisioterapii, 324p (in Russian)
Illarionov VE (2004) Scientific and practical foundations of information medicine. Moscow: Librocom, 184p (in Russian)
Ivanov EM, Zhuravskaya NS (2005) Major aspects of chronic bronchitis. Vladivostok: Dalnauka, 212p (in Russian)
Karepov GV (1991) Therapeutic exercise and physiotherapy in the system of rehabilitation of patients with traumatic spinal cord disease. Kiev: Zdorovya, 184p (in Russian)
Kipiani TI (1974) Mineral waters and functioning of the digestive system. Leningrad: Nauka, 58p (in Russian)
Klyachkin LM, Schegolkov AM (2000) Medical rehabilitation of patients with diseases of internal organs. A guide for doctors. Moscow: Meditsina, 325p (in Russian)
Knipovich NM (1938) Hydrology of the seas and brackish waters. Moscow-Leningrad: Pishchepromizdat (in Russian)
Kneipp S (1990) My hydrotherapy. Kiev: LAN, 176p (in Russian)
Konovalov AI (2013) The formation of nanoscale molecular ensembles in highly diluted aqueous solutions. Herald of the Russian Academy of Sciences (vestnik Rossiiskoi Akademii Nauk) 83(12):1076–1082 (in Russian)
Kozlova LV, Kozlov SA, Semenenko LA (2013) The basics of rehabilitation. Rostov-on-Don: Phoenix, 475p (in Russian)
Komarova LA, Egorova GI (1994) Combined methods of instrumental physiotherapy and balneothermotherapy. St. Petersburg: St. Petersburg Medical Academy of Postgraduate Studies, 219p (in Russian)
Kurlov MG, Sobkevich AI (1921) Practice of classification of Siberian healing mineral waters. Tomsk: Tomskiy fisioterapevticheskiy institute, 52p (in Russian)
Kulikov AG, Voronina DD (2017) Drinking mineral waters in treatment and rehabilitation: a modern view of the problem. Fizioterapiya. Bal'neologiya i Reabilitatsiya (russian Journal of Physiotherapy, Balneology and Rehabilitation) 3:116–120 (in Russian)
Kulakov VYa, Oransky IE (1970) Therapeutic mineral water. Sverdlovsk, 87p (in Russian)
Kulikov GV, Zhevlakov AV, Bondarenko SS (1991) Mineral medicinal waters of the USSR: Book of reference. Moscow: Nedra Publishers, 399p (in Russian)
Malkov AV, Pershin IM, Pomelyayko IS, Utkin VA, Korolyev BI, Dubogrey VF, Khmel VV, Pershin MI (2015) Kislovodsk deposit of acidulous waters: system analysis, diagnostics, forecast, management. Moscow: Nauka, 283p (in Russian)
Malyavin AG (2006) Respiratory medical rehabilitation. Moscow: Practicheskaya meditsina, 415p (in Russian)
Manshina NV (2007) Balneology for everyone: going to the resort for health. Moscow: Veche, 592p (in Russian)
Malkhazova SM, Ezerova LV, Kotova TV et al (2019) Faculty of geography. Moscow: Moscow State University, 304p
Maloletkina LA, Ulashchik VS (1999) Curative physiotherapeutic factors and clotting. Minsk: Nauka i tehnika, 118p (in Russian)
Mavritsky BF (ed.) (1985) Atlas of thermal water resources of the USSR (1:10000000). Moscow (in Russian)
Olefirenko VT (1989) Water treatment Moscow: Meditsina, 285p (in Russian)
Oransky IE (1988) Natural healing factors and biological rhythms. Moscow: Meditsina, 285p (in Russian)

Order of the Ministry of Health of the Russian Federation dated 07.06.18 (2018) No. 321n "On the approval of lists of medical indications and contraindications for spa treatment" (in Russian)
Ovchinnikov AM (1963) Mineral water (the study of mineral water deposits with
Ponomarenko GN, Chervinskaya AV, Konovalov SN (eds) (1998) Inhalation therapy. St. Petersburg: SLP, 233p (in Russian)
Sahakyan AG (1975) Spa treatment of colitis and rectal diseases. Stavropol: Stavropolskoe knizhnoye izdatelstvo, 159p (in Russian)
Samoilov OYa (1957) The structure of aqueous solutions of electrolytes and ion hydration. Moscow: USSR Academy of Sciences Publishing House, 179p (in Russian)
Semikhina LP (2005) An inductive method for determination of dielectric properties of liquids. Nauchnoe Priborostroenie (scientific Instrumentation) 15(3):83–87 (in Russian)
Shautsukova LK (ed.) (1976) Study on the mechanism of balneological factors' influence on regulatory systems of the body. Collection of scientific works. Nalchik, 164p (in Russian)
Shchukarev SA (1934) Endeavours to provide a general overview of Georgian waters from a geochemical point of view. Trudy Instituta Kurortologii 5:159–167 (in Russian)
Sheiko NI, Manshin NV (eds.) (2011) Caucasian Mineral Springs. Moscow: Veche, 304p (in Russian)
Shirokova VA (2010) Hydrochemistry in Russia: an outline of history. Moscow: Institute for the History of Science and Technology of the Russian Academy of Sciences (IHST RAS), 274 p (in Russian)
Shkolenko RL (ed.) (1973) Spa treatment of diseases of the digestive system and metabolism. Pyatigorsk: Stavropolskoe knizhnoe izdatelstvo (in Russian)
Smirnov-Kamensky EA (ed.) (1971) Therapeutic mineral waters. Pyatigorsk: PGNIIK, 279p (in Russian)
Smirnova IN (2004) Non-drug aerosol therapy of patients with chronic bronchitis and bronchial asthma. Extended abstract of MD Dissertation. Tomsk, 46p (in Russian)
Sosin IN, Buyavykh AG (2001) Physiotherapy of skin and sexually transmitted diseases. Simferopol, 268p (in Russian)
Starikov SM (2018) On medical geography of healing springs used in sanatorium organizations of the Ministry of Defense of Russia. Voyenno-Meditsinskiy Zhurnal 339(1):94–96 (in Russian)
Strelkova NI (1991) Physical methods of treatment in neurology. Moscow: Meditsina, 272p (in Russian)
Starikov SM, Polyaev BA, Bolotov DD (2013) Physical rehabilitation in complex treatment of patients with dorsopathy. Moscow: Krasnaya Zvezda, 154p (in Russian)
Sulin VA (1946) Oil-field waters in the natural water system. Moscow-Leningrad: Gostoptekhizdat, 96p (in Russian)
Vernadsky VI (1933) History of natural waters. Part 1. In: History of minerals in the Earth's crust. Vol. 2. Leningrad: Goshimtehizdat, 202p (in Russian)
Veselovsky VP, Popelyansky AYa, Sakhovsky PI, Khabriev RU (1982) Rehabilitation of patients with vertebrogenic diseases of the nervous system. Training manual for doctors-probationers. Leningrad: Kazan Institute of Postgraduate Medical Training, 48p (in Russian)
Vinokurov BL (2008) Natural hydrocarbonate sodium-calcium acidulated arsenous boron mineral waters of Sochi recreation zone Krasnaya Polyana. Sochi: RIO NIC kurortologii and reabilitatsii Roszdrava, 148p (in Russian)
Voronov AN, Vinograd NA (2004) The basic principles for creating a modern classification of mineral and medicinal waters. Vestnik of Saint Petersburg University Ser. 7(2):94–96 (in Russian)
Voronina EYu, Zelinskaya EV (2009) Theoretical aspects of the use of hydro-mineral raw materials. Moscow: Akademija Estestvoznanija, 118p (in Russian)
Vorobiev MG, Parfenov AP (1982) Physiotherapy and balneology. Leningrad: Military Medical Academy, 248p (in Russian)
Volkov VS, Tsikulin AV (1989) Treatment and rehabilitation of patients with hypertensive desiase in a clinic. Moscow: Meditsina, 254p (in Russian)
Zernov SA (1934) General hydrobiology. Moscow-Leningrad: Biomedgiz, 158p (in Russian)

Chapter 2
Mud Springs (Peloids)

Abstract The chapter is devoted to the types, distribution, and application of mud spring. Depending on their physicochemical properties and formation conditions, the description of medicinal mud includes sulfide silt mud, sapropel, peat mud, and mud volcanic breccia. Deposits of medicinal mud in Russia are shown on the series of maps.

Keyword Medicinal mud · Peloids · Deposit of mud · Sapropel · Sulfide silt mud · Peat mud · Mud volcanic breccia · Distribution of mud spring · Russia

2.1 Mud Springs: Types, Distribution, and Application

Medicinal mud (or peloids) is a natural colloidal organomineral formation having high plasticity, high effective heat capacity, and slow heat transfer, containing biologically active substances (salts, gases, vitamins, enzymes, hormones, etc.) and living microorganisms. The most important properties of mud that determine its biological activity and medicinal properties are viscosity, heat capacity, pH, Eh (redox potential), content of hydrogen sulfide, silicic acid, trace elements, specific and non-specific microflora, organic matter, etc. According to the WHO classification, peloids are natural resources consisting of natural organic and mineral substances and microorganisms forming mainly under anaerobic conditions near salt and fresh water bodies (Application of Peloid Therapy… 2015).

The term "medicinal mud" includes sediments of various reservoirs, peat deposits of swamps, eruptions of mud volcanoes, and other natural formations consisting of water, mineral, and organic substances in the form of a homogeneous finely dispersed plastic mass formed under the influence of geological, climatic, hydrogeological, biological, and other natural factors (Fig. 2.1). Correspondingly, only that special mud acquires healing properties, which nature "prepared" painstakingly for centuries. Microorganisms are the first of mud components. It has been shown that one tablespoon of mud may contain more than one billion microorganisms, each of which contributed to the enrichment of this natural suspension with biogenic components and new compounds, which led to the high therapeutic value of peloids (Muradov 2007; Badalov and Krikorova 2012).

S. Malkhazova et al., *Healing Springs of Russia*,
https://doi.org/10.1007/978-3-030-83534-7_2

Fig. 2.1 Mud of Kirek Lake (Tomsk Oblast)

A characteristic feature of peloids is high microbiological activity, distinguishing them from other similar natural formations. The high activity of bacteria, fungi, and other organisms promotes the decomposition of organic and animal residues, and enriches medicinal mud with humic substances and bitumen; it produces ammonium ion, hydrogen sulfide, carbon dioxide, and methane. A continuous flow of organic substances from the aquatic environment to the sediment ensures the constant activity of microorganisms that maintain a stable content of such important and unstable organic compounds as vitamins, enzymes, and hormones in the mud (Muradov 2007; Trebuhov and Krikorova 2009).

Biogenic elements are often concentrated in medicinal mud. Their content is small (less than 1% in dry mud), but their role in the biochemical processes of peloid formation is huge. For example, in these healing springs cultivated by nature for so long, there are a large number of elements that are important for human health: iodine, potassium, calcium, magnesium, iron, bromine, sodium, sulfur, nitrogen, phosphorus, silica, etc. Calcium strengthens the skeleton and cardiovascular activity, iron helps the body's circulatory system function, bromine has a beneficial effect on the nervous system, potassium plays an important role in metabolic processes, magnesium prevents muscle and vascular spasms, and sulfur actively fights fungal skin diseases, while sodium deficiency leads to fatigue.

Depending on their physicochemical properties, the classification of medicinal mud adopted in Russia and compiled with consideration of their formation conditions, includes sulfide silt mud, sapropel, peat mud, and mud volcanic breccia (Application of Peloid Therapy… 2015). Deposits of medicinal mud in the Russian Federation are shown on Map 2.1.

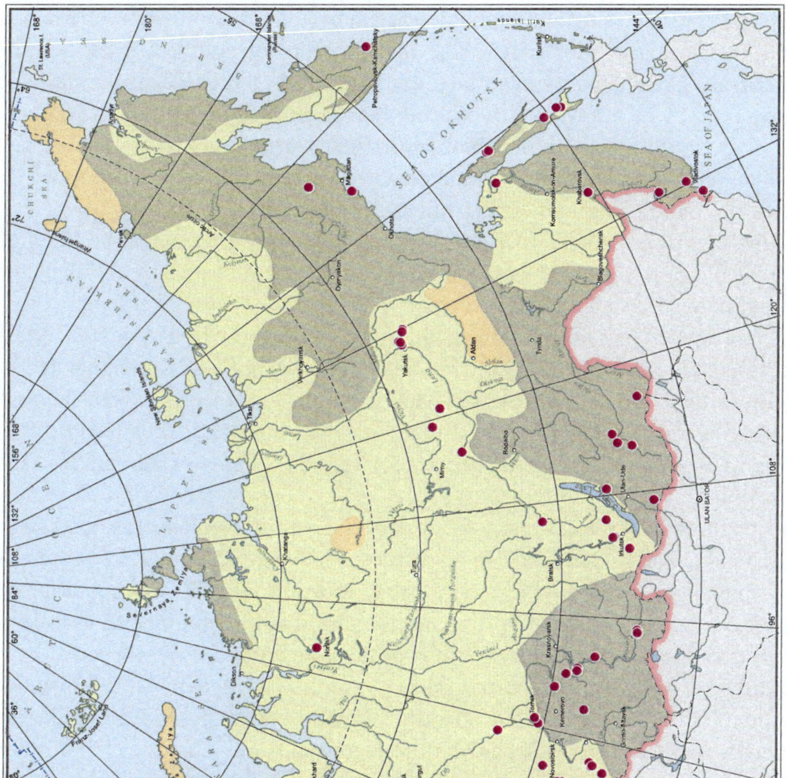

Map 2.1 Treatment mud deposits

Sulfide silt mud is the bottom sediment of predominantly mineralized water bodies, poor in organic matter and enriched with iron sulfides and water-soluble salts (Map 2.2). Sulfide silt mud is characterized by the predominance of mineral components over organic substances. The main feature of sulfide silt mud is the content of various gases (hydrogen sulfide, methane, carbon dioxide, ammonia) and organic substances, which have a pronounced therapeutic effect. Among the organic substances with important antimicrobial properties, various acids, pigments, and penicillin-like substances can be distinguished. These compounds are produced by various bacteria, mold, fungi, and antagonist actinomycetes.

The following groups can be distinguished among sulfide silt mud, depending on the genesis of the deposit: continental mud of salt lakes, mud of coastal lakes and estuaries, and mud of lake and spring origin.

Continental mud of salt lakes includes the mud of Bolshoy Tambukan and Maly Tambukan Lakes, the Small and Big Salt Lakes in Stavropol and Krasnodar Krais, Karachi (Novosibirsk Oblast), Yarovoye (Altai Republic), and Uchum (Krasnoyarsk Krai). There is a lot of silt sulfide mud in Crimea: in Saki and Yevpatoria (Otar-Moinak Lake), as well as in Kipchak, Krasnoe, Terekly, Olenye, Uzunlarskoe, and Chokrak Lakes.

The most famous sources of *mud of coastal lakes and estuaries* are located in Anapa (Krasnodar Krai), Sadgorod (a suburb of Vladivostok), and Saki (Crimea). The Kiziltash estuary located within the Anapa resort provides its raw materials to the resorts of Sochi, Gelendzhik, Tuapse, etc. In addition, the mud of this estuary is also packaged and exported to other regions of the country and even abroad.

Sulfide silt mud of lake and spring origin is silt sediment of open reservoirs with a large amount of sulfate and high hydrotroillite content. On the base of these healing springs, there are such health resorts as Marcial Waters in Karelia, Sergievsky Mineral Waters in Samara Oblast, Khilovo in Pskov Oblast, Solvychegodsk in Arkhangelsk Oblast, Krasnousolsk in the Republic of Bashkortostan, Ust-Kut in Perm Krai, the Paratunsky resort of Kamchatka Krai, and resorts in Staraya Russa (Novgorod Oblast).

Sapropels (from the Greek. *sapros*–rotten, *pelos*–silt, clay) are organogenic bottom sediments of mostly freshwater bodies (Map 2.3). Their formation is associated with the microbiological decomposition of algae and other plant and animal residues on the lake bottom. The healing mud layer at the bottom of the reservoir can reach more than 10 m in thickness, but usually, only the upper 1–2 m are developed. Sapropel has diverse colors: bluish, pink, or brown.

The therapeutic value of these deposits is determined both by their high thermal properties, and by the presence of organic substances and biostimulants: vitamins, enzymes, and hormones. Sapropelic mud has good ductility, viscosity, and stickiness; and contains a large number of chemical elements (cobalt, manganese, copper, zinc, boron, molybdenum, iodine, bromine, etc.) and vitamins (B, C, D, riboflavin, folic acid, etc.) (Kilina 1997).

Sapropels in Russia are used in many resorts: at Moltaevo Lake in Sverdlovsk Oblast; at Podbornoe Lake, Akachkul Lake and Bolyash Lake in Chelyabinsk Oblast,

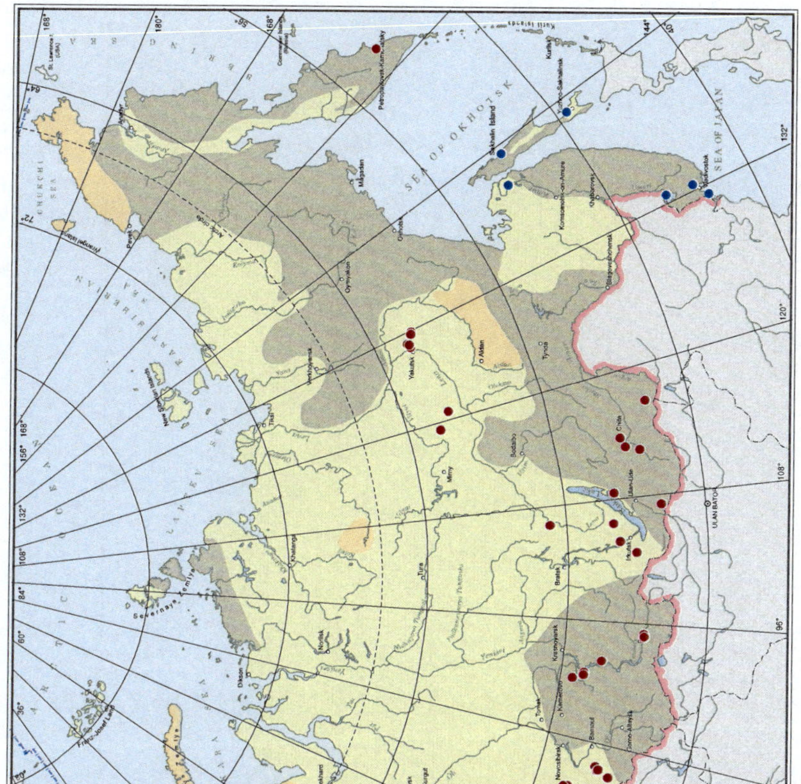

Map 2.2 Sulfide silt mud

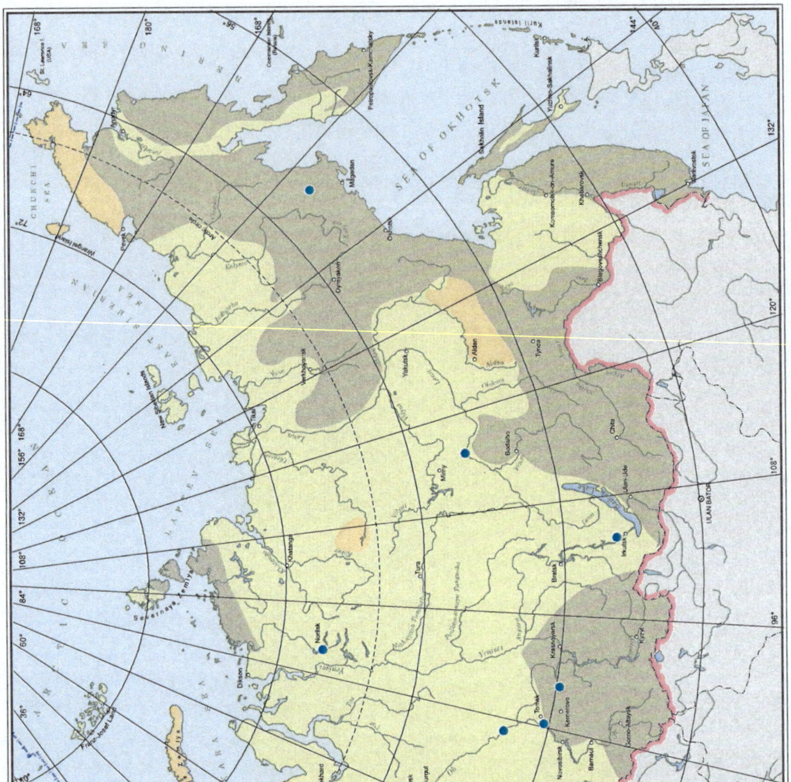

Map 2.3 Sapropels

at Mutnoye Lake in Smolensk Oblast, in Yaroslavl Oblast, Novgorod Oblast, Tyumen Oblast, and some other regions. Sapropelic mud of the Sestroretsk deposit (a resort area of St. Petersburg) has unique physical and chemical properties. These are buried sapropels of the ancient Litorin Sea, called "gittia clays."

Map 2.4 shows **peat mud and mud volcanic breccia** deposits. *Peat* is an organogenic sediment of low and high bogs, widespread in various natural areas of the country. This healing mud is formed under conditions of waterlogging, which caused poor oxygen supply, and with the active participation of anaerobic microorganisms. The mud of low bogs is characterized by a higher content of organic matter. The therapeutic value of peat is due to its high thermal properties (high ability to retain heat) and large content of therapeutically active organic substances: humic acids, lipids, and bitumen. Since bog sediments are characterized by an acidic reaction and high bactericidal activity, they have an anti-inflammatory effect on the human body. They stimulate enzymatic activity and accelerate regeneration processes. It has been shown that peat mud treatment has a complex effect on basic body functions: blood circulation, respiration, metabolism, etc. Peat mud is successfully used in the resorts of the Caucasian Mineral Waters, Urals, Siberia, the Far East, and other regions of Russia. Typical peat mud deposits are Goreloye Lake in the Kaliningrad Oblast (resorts in Svetlogorsk, Otradnoye, and Zelenogradsk), Ryabtsevskoye Lake in the Tver Oblast (Kashin resort), Dvurechenskoye Lake in the Lipetsk Oblast (Lipetsk resort and others).

In addition, there is unique volcanic-type mud — *mud volcanic breccia*, common in areas of oil and gas manifestations and having a deep origin. These semi-liquid clay formations erupt to the surface under the pressure of gases and groundwater, but as they move, they are enriched with gases, humic and fulvic acids, microelements (lithium, barium, manganese, strontium), and high concentrations of iodine and boron. Large concentrations of bromine in mud volcanic breccia cause it to have a sedative and anti-inflammatory effect. In the Russian Federation, this mud is found on the Taman and Kerch Peninsulas, as well as on southern Sakhalin. At present, mud volcanic breccia is used in our country in limited quantities in Anapa (Azov Hill). In the near abroad, Azerbaijan (Apsheron, Prikurinsky Lowland) is the largest area of mud volcanic breccia spreading, where there are more than 250 sources.

In addition, some classifications also distinguish *hydrothermal mud*, the genesis of which is also associated with tectonics. This mud develops in areas with increased volcanic activity (Kamchatka, Kuril Islands) as a result of leaching and decomposition of volcanic rocks by high-temperature gas-steam jets containing carbon dioxide and hydrogen sulfide. Hydrothermal therapeutic mud in Kamchatka is associated with the so-called "mud pots" found in the Kronotsky Reserve (Valley of Geysers), at the Nalychevo thermal springs, at the Koshelevsky volcano, etc. The remoteness of these springs and the absence of settlements near them complicate the study of their physicochemical and biological properties, and they are practically not used in balneology (Muradov 2007).

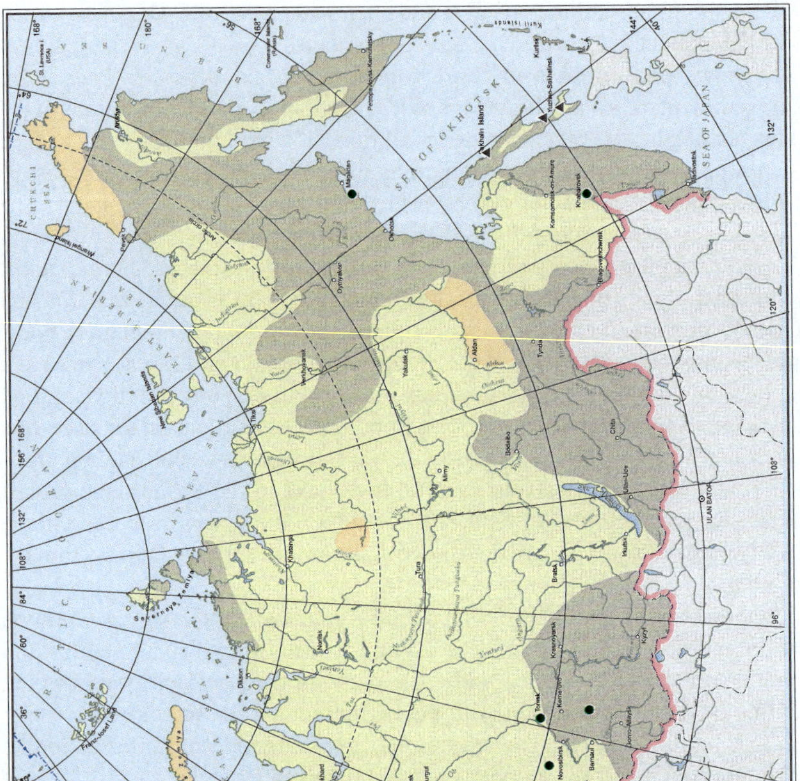

Map 2.4 Peat mud and mud volcanic breccia

2.2 The Use of Mud Springs

Many mud treatment methods have been developed and are currently being applied. The therapeutic effect of mud consists of thermal, mechanical, and chemical effects. The effect of peloids can be various and depends on both the combination of the properties of applied mud and the condition of the patient.

As noted previously, although all mud has a high heat-holding ability, their thermal properties can be different: for example, the thermal conductivity of sulfide silt mud is almost double that of peat. The therapeutic effect also depends on the measure of acidity or alkalinity (pH value) and degree of substrate mineralization (Ponomarenko 2013).

The therapeutic effect of peloids is also based on their chemical composition and the presence of biologically active substances. A theory of the mechanism of therapeutic mud's chemical effect on the human body was developed by the founder of Saki resort, Dr. N. A. Auger, who hypothesized that "something enters the body from the mud" in 1827. In the middle of the twentieth century, this assumption was confirmed by numerous studies by scientists (Ponomarenko 2013). Firstly, as a result of microorganism activity, in the silt fractions of water bodies, a so-called "volatile complex" is formed, consisting of hydrogen sulfide and other gases that can penetrate the skin. Secondly, there is a whole complex of organic acids in therapeutic mud that act on the skin.

Biologically active substances and microelements have also been found in therapeutic mud. The body's reaction upon the application of mud to the skin is considered to have two phases: neuro-reflexive and neuro-chemical. For the first phase, direct irritation of skin nerve receptors is characteristic, which leads to excitation of corresponding nervous centers, and, in the final result, to the reorganization of the body's metabolism. The second phase is characterized by the formation of biochemically active substances in the skin (under the influence of compounds that have penetrated from mud), which, when entering the bloodstream, cause a cascade reaction. It has been proved that peloids having, among other things, antioxidant properties, have both analgesic and anti-inflammatory effects, increase blood circulation, and improve metabolism. During mud procedures, pulse and respiration become rates increase, blood pressure rises and then decreases, sweating intensifies, and the urinary function increases. The autonomic nervous system plays a significant role in the development and course of these changes. In addition, peloids have a pronounced antimicrobial effect, which is explained by a high content of sulfide groups, bromine ions, zinc, and antibiotic analogues. Therapeutic mud of the Sestroretsk deposit has fungicidal (antifungal) properties (this therapeutic mud belongs to the group of sulfur-glandular sapropels, or gittia clays) (Trebuhov 2007).

Indications for mud therapy are quite extensive: first of all, diseases of the musculoskeletal system and connective tissue (arthritis; polyarthritis of rheumatic, infectious, toxic, metabolic, endocrine, or traumatic origin, etc.) (category XII); in addition, mud therapy is indicated for diseases of the nervous system (radiculitis; plexitis; neuritis; infectious, rheumatic and secondary neurofibromyositis; consequences of injuries and other traumas of the peripheral nervous system that do not require surgical treatment, etc.) (category VI); diseases of the respiratory system (chronic bronchitis; infiltrates and adhesions in the thorax cavity after injuries, operations, infections (excluding tuberculosis)) (category X); diseases of the digestive system (gastric and peptic ulcer disease without exacerbation, diseases of the stomach operated due to ulcers, chronic colitis, chronic hepatitis, chronic cholecystitis, pericholecystitis, etc.) (category XI); diseases of the genitourinary system (chronic endometritis, metritis, parametritis, adnexitis with underlying unchanged hormonal function or hypestrogenism, chronic prostatitis, gynecological diseases) (category XIV); diseases of the skin and subcutaneous tissue (consequences of injuries to bones and soft tissues, long-term healing wounds, trophic ulcers, keloid scars) (category XII) (Ponomarenko 2013).

Contraindications for mud therapy include oncological diseases (at any stage), benign neoplasms (fibromyomas, mastopathies) (category II); tuberculosis in any location (category IX); acute inflammatory processes and exacerbations of chronic diseases (category II); some cardiovascular diseases, and pregnancy. Therefore, before mud therapy, a specialized medical consultation is necessary.

References

Application of Peloid Therapy in treatment and rehabilitation programs. Clinical recommendations. Approved at the XIII International Congress "Rehabilitation and spa treatment". Moscow. 2015 (in Russian)

Badalov NG, Krikorova SA (2012) Mud therapy: theory, practice, problems and development prospects. Voprosy kurortologii, fizioterapii i lechebnoi fizicheskoi kul'tury (Problems of Balneology, Physiotherapy, and Exercise Therapy) 3:50–54 (in Russian)

Kilina ES (1997) Biological activity of therapeutic sapropelic mud of Siberia. Voprosy kurortologii, fizioterapii i lechebnoi fizicheskoi kul'tury (Problems of Balneology, Physiotherapy, and Exercise Therapy) 2:23–25 (in Russian)

Muradov SV (2007) Ecological solution to the problems of modern mud therapy. Petropavlovsk-Kamchatsky: Vitus Bering Kamchatka State University, 266p (in Russian)

Ponomarenko GN (ed) (2013) Physiotherapy: a national guide. Moscow: GEOTAR-Media, 408p (in Russian)

Trebuhov YA (2007) Composition characteristics of natural medicinal clays. Voprosy kurortologii, fizioterapii i lechebnoi fizicheskoi kul'tury (Problems of Balneology, Physiotherapy, and Exercise Therapy) 2:36–38 (in Russian)

Trebuhov YA, Krikorova SA (2009) Medicinal muds and clays. Pelotherapy and other types of thermotherapy. In: Razumov AN, Bobrovnitsky IP, Vasilenko AM (eds) The textbook on rehabilitation medicine. Vosstanovitelnaya meditsina, Moscow, pp 114–117 (in Russian)

Chapter 3
Healing Springs in Regions

Abstract In this chapter, the characteristics and original maps of healing springs sites in some regions of Russia are given. The Caucasus represents the world-famous healing springs and occupies a special place due to its nature. The Transbaikal region is one of the largest regions of global significance in its reserves and potential resources of mineral water; it occupies one of the leading places in Russia, second only to the Caucasus. The Kamchatka Peninsula is characterized by the significant originality of its healing resources of thermal water, which formed under unique natural conditions. Less well-known Tyva Republic has a unique geographical location, at the junction of South Siberian taiga forests and Central Asian steppes and deserts. The phenomenon of arzhan treatment (healing springs), which has been an integral part of Tyvan culture and traditions since ancient times, is unique.

Keywords Healing spring · Caucasus · Transbaikal · Kamchatka · Tyva

In this section, for an example, based on data collected by authors during expeditions and provided by regional departments, characteristics and original maps of sites are given of healing springs in some regions of Russia: Caucasus, Crimea, Tyva, Transbaikal, and Kamchatka (Figs. 3.1 and 3.2).

3.1 Caucasus

Among world-famous territories, the Caucasus occupies a special place due to its nature. Natural springs of healing water have long been known here (Map 3.1). In 1793, the Russian Academy of Sciences sent Academician Peter Simon Pallas to the Caucasus with an order to describe the Caucasian springs in detail. Scientific data on Caucasian mineral water appeared, at that time and the gradual development of the region began. Excellent local conditions contribute to the popularity of the mineral water (Kotlyakov et al. 2017).

S. Malkhazova et al., *Healing Springs of Russia*,
https://doi.org/10.1007/978-3-030-83534-7_3

Fig. 3.1 Sochi. Well room of Matsesta gallery

Fig. 3.2 Shira Lake in Khakassia. The main therapeutic factors are mineral water of the lake and well, silt mud, and steppe climate

3.1.1 Stavropol Krai

This territory is home to the famous Caucasian Mineral Waters resort (CMW), where treatment is based on the use of various types of natural mineral water, treatment mud, and landscape and climatic conditions of the region.

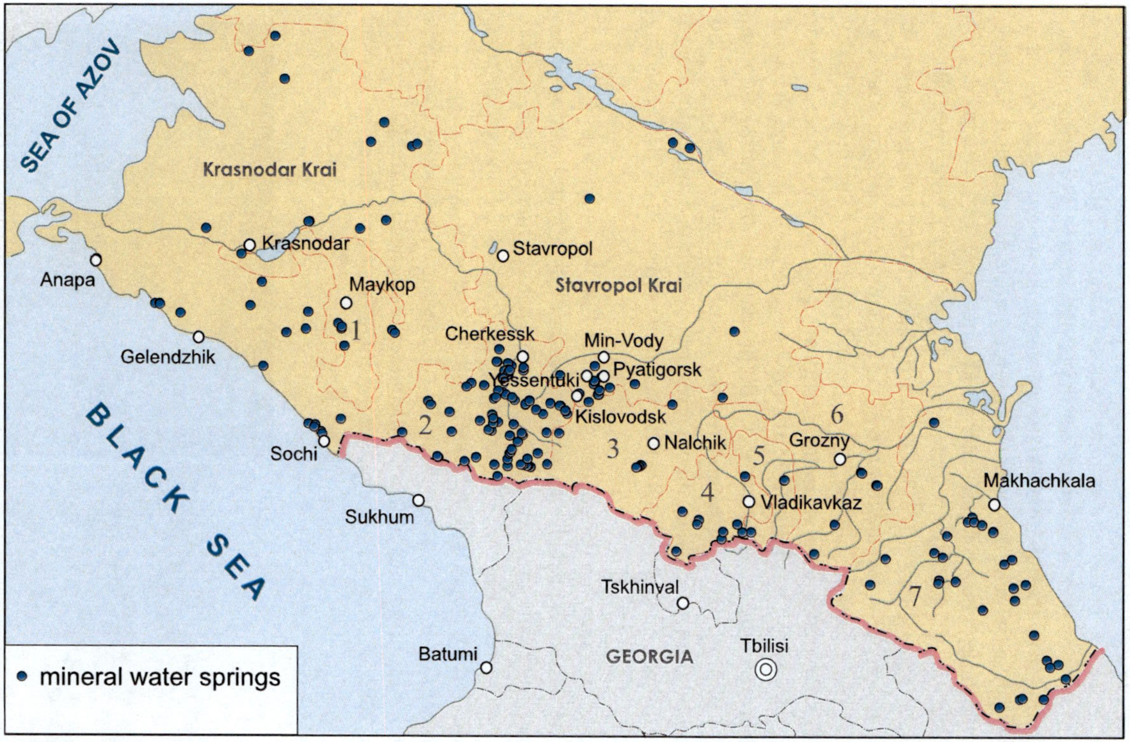

1. Republic of Adygea
2. Karachay-Cherkess Republic
3. Kabardino-Balkar Republic
4. Republic of North Ossetia (Alania)
5. Republic of Ingushetia
6. Chechen Republic
7. Republic of Dagestan

МАСШТАБ 1 : 5 000 000

Map 3.1 Mineral water springs of Caucasus

Currently, 20 deposits of various types of mineral water, with a total reserve of more than 16,000 m^3/day, have been explored and developed on the territory of the CMW region. However, only mineral water from Kislovodsk, Essentuki, Beshtaugorsk, Pyatigorsk, Zheleznovodsk, and Kumagorsk deposits, with a total reserve of about 7,000 m^3/day, is used in treatment at the resorts. The remaining deposits are partially used for industrial bottling of mineral water (Schaefer 2017).

Kislovodsk resort is located in the south of the CMW region. For drinking and balneological treatment at the Kislovodsk resort, carbonic mineral water is used from the same deposit, at which four production sites are located: Central, Olkhov, Berezov, and Podkumsk. The mineral water production reserve is 10 m^3/day of sulfate Narzan, 55 m^3/day of dolomite Narzan, and 1,470 m^3/day of combined Narzan produced in the Central section (Fig. 3.3). The carbonated mineral water reserve of the Olkhov site is 164 m^3/day, and the reserve of the Podkumsk site is 95 m^3/day. For drinking treatment, the resort uses sulfate, dolomite and combined Narzan yielded by wells from a depth of more than 100 m, and for balneotherapy, mainly mineral water from the Narzan spring and partially from the Olkhov site wells is used. Mineral water from the Berezov site, with a reserve of 300 m^3/day, and the Kumsk deposit, with a reserve of 1,335 m^3/day, is diverted from the resort mineral water supply and used for industrial bottling only. Thus, the total reserve of carbonic mineral water of the Kislovodsk resort is 1,794 m^3/day, which is currently used by 30–50%, depending on the season.

Essentuki Resort is located in the central part of the CMW region (Fig. 3.4). The resort's hydromineral base is provided by mineral water of the Central section of the Essentuki deposit, where the medicinal and medicinal-table water types Essentuki No. 17 and Essentuki No. 4 are developed, with reserves of 22 and 39 m^3/day, respectively, as well as hydrogen sulfide water, with a reserve of 200 m^3/day.

Thermal nitrogenous and carbonic water is yielded in the Middle-Essentuki section. A mixture of nitrogenous and carbonic water is supplied to the balneology facilities through a six-kilometer stainless steel mineral water pipeline and is used

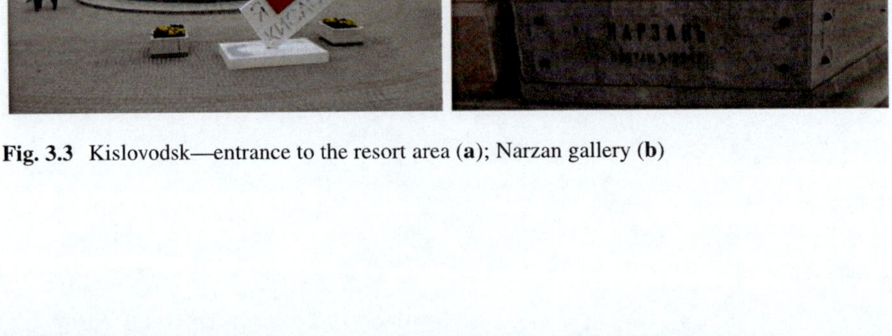

Fig. 3.3 Kislovodsk—entrance to the resort area (**a**); Narzan gallery (**b**)

Fig. 3.4 Essentuki. The Institution for Mud Cures named after N.A. Semashko

in the resort under the name Essentuki Novaya for drinking and balneological treatment. In addition, carbonic thermal water of the Central Beshtaugorsk section of the Beshtaugorsk deposit, with a reserve of 490 m^3/day, is used to fully provide for the resort. It is also transported to the resort via stainless mineral water pipeline longer than 13 km long and is used for drinking and balneological treatment. Mineral water of the Eorn, Buguntinsk, Novo-Blagodarnensk, Bykogorsk, and Zapadno-Bykogorsk sections of the Essentuki deposit, with a total reserve of 910 m^3/day, is not used at the resort; it is used only for industrial bottling of mineral water such as Essentuki No. 17, Essentuki No. 4, Mountain Essentuki, etc. Thus, the real hydromineral base of the resort is 751 m^3/day of various mineral water types.

Pyatigorsk resort is located in the eastern part of the CMW region. It is the most well provided for hydromineral resource among the CMW resorts. The operational reserve of mineral water is the following:

- carbonic-sulfide water –1,740 m^3/day,
- carbonic water (Narzan) –357 m^3/day,
- salt-alkaline water –6.6 m^3/day,
- radon water –400 m^3/day, and
- radon water from well No. 113 of the Beshtau mountain –306.2 m^3/day.

Total: 2,809.8 m^3/day.

The available mineral water reserve is currently being used by only 7–10%.

Zheleznovodsk resort is located in the northeastern part of the CMW region (Fig. 3.5). The hydromineral base of the resort is the mineral water of the Zheleznovodsk deposit, on which three types of mineral water have been explored and produced:

Type 1 — low-mineralized sulfate-hydrogencarbonate and calcium-sodium mineral water of the main Zheleznovodsk type, differing in temperature: hot (37–62 °C), warm (20–37 °C), and cold (up to 20 °C). They are capped by springs and wells in Upper Cretaceous and Elburgan rocks, as well as in sediments of the Goryachiy Klyuch formation. Hot springs include the Slavyanovsky, Smirnovsky No. 1 and 2, Semashko, and hot Nelyubinsky water springs, as well as the No. 69, 1-Yuzhnaya, and 59 wells, etc. Warm water is capped by the Vladimir spring and well No. 64. Coldwater is capped by the Gaazovsky, cold Nelyubinsky, Zavadovsky, and Nezlobinsky springs, and by well No. 54. Since the resort's foundation, the water of the first type has been used for drinking and balneological treatment, as well as for bottling as Slavyanovskaya and Smirnovskaya. Its operating reserve is 1,176 m^3/day.

Type 2 includes carbonic, high-temperature, low- and medium-mineralized hydrogencarbonate -sulfate and calcium-sodium water capped by well No. 70 from Lower Cretaceous deposits at a depth of more than 1,100 m. Its operating reserve is 700 m^3/day. There were plans to use this type of water for drinking and balneotherapy, as well as for industrial bottling. At least for now, this water is on standby.

Type 3 includes cold low- and medium-mineralized, hydrogencarbonate-chloride sodium water capped by wells No. 65, 61, and 63 from sediments of the Goryachy

Fig. 3.5 Zheleznovodsk. Pushkin gallery

Klyuch formation. Its operating reserve is 10 m^3/day. Mineral water of this type has been used for drinking treatment only. Currently, it is not used at the resort. Its available hydromineral resource is utilized by no more than 30–40%.

Kumagorsky Regional Hospital of Rehabilitation Treatment (Kumagorsky resort) is based on mineral water from the eponymous deposit. In its chemistry, Kumagorsk mineral water is close to the sulfide water of the Goryachy Klyuch resort in the Krasnodar Krai. It is successfully used for balneotherapy. Its operating reserve is 260 m^3/day. Its current use does not exceed 20%.

3.1.2 Krasnodar Krai

Krasnodar Krai is the leading region of Russia for recreational-tourist, spa resort, and balneological services. This is mainly due to the presence of an extensive network of mineral water springs (Fig. 3.6) and treatment mud, which, combined with unique climatic and natural-recreational factors, creates a specific system of recreational and resort complexes (Nagalevsky et al. 2014; Mironov and Khodasevich 2018).

The resource base of the regional balneological complex consists of 58 mineral water deposits (Fig. 3.7), with a total reserve of 23,670 m^3/day. (Nagalevsky et al. 2014). There are 74 sections of mineral water deposits that have been explored; the deposits in use are the following:

(1) in the northern zone: Yeisk, Privolnoe as a source of drinking water,
(2) in the central zone: Krasnodar, a section of medicinal and drinking water,
(3) in the foothill zone: Chernomorie, Psekupsky (left-bank and right-bank sections), Fanagoria, Labinsk (medicinal water), Labinsk (thermal and medicinal-table water), Khadyzhensk (medicinal-table water),
(4) on the Black Sea coast: Dzhemetyn, Pioneer, Anapa, Semigorsk, Divnomorskoye, Plastun, Sochi (Mamai site), Sochi (Matsesta site), Sochi (Kudepsta site), Chizhepse, Krasnaya Polyana.

Fig. 3.6 Drinking well room for ingesting mixed alkaline chloride-hydrogencarbonate sodium water with a high content of organic substances. Krasnodar municipal hydropathical establishment

Fig. 3.7 Krasnodar Krai. Labinsky district, Laba resort, open-air drinking well room

Unfortunately, the potential of the local mineral water has been underestimated and it is poorly used in industrial bottling (Nagalevsky et al. 2014):

- water of the Khadyzhensk deposit is comparable in quality to the mineral water of Essentuki and Borjomi (production is 7% of the possible);
- water of the Anapa deposit is similar to the mineral water of Uglich, Mirgorod, and Feodosiya (production is 2.5% of the possible);
- medicinal water of the Semigorsk and Velikovechnoe deposits is generally unique and has no analogues of its health-improving properties (production is 1.8% of the possible).

In the region, there are deposits of mineral water of various hydrochemical classes (Nagalevsky et al. 2014; Mironov and Khodasevich 2018):

- sulfide mineral water is used for therapeutic purposes at the resorts of Sochi, Yeysk, Goryachiy Klyuch
- iodine-bromine mineral water is present in 20 hydropathical establishments in the region (Krasnodar, Sochi, Goryachiy Klyuch, Khadyzhensk, Anapa, Neftyanaya, Apsheronsk, Labinsk, Velikovechnoe, and others)
- carbonic arsenic mineral water is yielded from the Chvizhepse spring;
- alkaline mineral water is used at the Goryachiy Klyuch, Khadyzhensk, and Gelendzhik resorts; Goryachiy Klyuch water with low and medium mineralization is bottled industrially;
- there are rare types of mineral water that are unique in their chemistry (for example, the water of deposits at Spokoynaya station);

- ferruginous slightly alkaline springs in the Seversk district are known as Zaporozhye springs.

3.1.3 Republic of Adygea

Republic of Adygea has unique natural resources, including mineral and thermal springs. In Adygea, four mineral water deposits with approved operational reserve of 6,007 m^3/day are in use. (Korinevich 2015; Nechiporova 2017).

1. *Khan deposit.* Low-mineralized (1.5–1.8 g/l), sodium hydrogencarbonate, low thermal (22–28 °C) water with a high content of organic compounds. Medicinal-table water is bottled under the Maykop brand. Confirmed operational reserve of 210 m^3/day.
2. *Abadzekhskaya deposit.* Medium-mineralized (5.4–6.4 g/l), sulfate–chloride sodium sulfide with medium concentration (60–95 mg/l), high thermal (57–62 °C) water. Confirmed operational reserve of 172 m^3/day.
3. *Maykop deposit.* Highly mineralized (23 g/l) sodium chloride water with a high content of iodine (22–27 mg/l) and bromine (90–130 mg/l). It is thermal water (60–68 °C). Confirmed operational reserve of 600 m^3/day.
4. *Kurdzhip deposit.* Low-mineralized (1.3–4.0 g/l) hydrogencarbonate -chloride sodium water with a high content of organic substances, slightly thermal (32–33 °C) medicinal-table water. It is bottled under the name Caucasian Spring. Confirmed operational reserve of 250 m^3/day. In explored deposits, mineral water has been discovered at 14 wells with a depth of 400 to 2,000 m. There are seven wells in commercial development. A number of medical institutions of the republic operate on the basis of mineral and thermal deposits. Based on exploration work carried out in 2009, nine new sites were identified in the valleys of the Belaya, Fars, and Hokodz Rivers. The resource potential of the mineral water at the sites explored is 3,500 m^3/day (Korinevich 2015).

3.1.4 Karachay-Cherkess Republic

Climatic conditions, a diverse landscape with divisions into four geomorphological units (highlands, midlands, lowlands, and plains), and the presence of water bodies are important components of the natural resort wealth of the republic. A special place among natural resources of Karachay-Cherkessia belongs to mineral springs. Down through the ages, the locals have been aware of the healing properties of carbonic water and have used it to heal all kinds of ailments: musculoskeletal, gastrointestinal, cardiovascular diseases, and a number of others (Vasin et al. 2015).

The quantity and variety of mineral water in Karachay-Cherkessia is compared to the Caucasian Mineral Waters region. Natural mineral springs, which number more than 200, have long been known here. Some of them are suitable for bottling and balneological applications. Currently, 12 mineral water deposits have been explored in the republic: the southern section of the Cherkess deposit, Verkhnepodkumsk, Teberda, Dombay, Kumsk, Chapaevsk, Vazhnensk, Karachaev, Krasnaya Gorka, Adyge-Khabl, Eshkakon, and Khabez deposits (Vasin et al. 2015). Their water is diverse in mineralization level and chemistry. The water of the Verkhnekumsk deposit is carbonic hydrogencarbonate -sulfate calcium-magnesium, with mineralization of 1.7 g/l; it is an analogue of Kislovodsk dolomite Narzan. Water of the Kumsk deposit is carbonic hydrogencarbonate -sulfate sodium-calcium, with mineralization of 1.7 g/l; it is an analogue of water of the well-known Kislovodsk and Darasun deposits. Water of the Teberda deposit is carbonic, hydrogencarbonate -chloride sodium-calcium, with mineralization of 6.1 g/l. Water of the Chapaevsk deposit is chloride-sulfate-sodium hydrogencarbonate, with mineralization of 3.6 g/l; its physicochemical composition is close to Feodosia-type water. Water of the Cherkessk deposit is iodine-bromine chloride sodium, with mineralization of 52.9 g/l. It resembles mineral water of the Semigorye, Naphtalan, Chartak, Razdan deposits. Water of the Vazhnensk deposit is carbonic chloride-hydrogen-carbonate sodium-calcium, with mineralization of 3.2 g/l. Its physicochemical properties are similar to that of Lugansk-type medicinal-table mineral water. Water of the Psyzh deposit is chloride-sulfate sodium-magnesium, with mineralization of 2.4 g/l; it is similar to Makhachkala-type medicinal-table water. Water of the Khabez deposit is chloride-sulfate sodium-magnesium, with mineralization of 2.5 g/l. Water of the Sary-Tyuz deposit is carbonic, hydrogencarbonate -sulfate, calcium-magnesium, with mineralization of 1.2 g/l; it is similar to Kishinev-type medicinal-table mineral water. Water of the Karachaev deposit is sodium chloride-hydrogencarbonate, with mineralization of 2 g/l; its physicochemical properties are similar to those of the well-known Aivazovsky-type medicinal-table mineral water. The Arkhyz deposit is represented by water with low mineralization (0.2–0.3 g/l). It is used by large bottling plants for sale as drinking table water.

3.1.5 Kabardino-Balkar Republic

In the Kabardino-Balkar Republic, there are more than 180 mineral springs and more than 100 lakes (Vasin et al. 2016), of among which are the unique Tambukan (therapeutic mud deposit) and Nizhnee Goluboye with a depth of 254 m and an all-year stable water temperature of 9.3 °C. The republic is rich in mineral water with valuable medicinal properties. It has carbonic, hydrogen sulfide, nitrogen subthermal, chloride-sodium, iodine-bromide chloride-sodium water with a high content of silicic acid and organic substances, hydrogencarbonate -chloride sodium water, etc. Even in ancient times, locals knew the healing properties of mineral springs and used them. This is evidenced by remains of primitive wooden and stone bathing devices found near mineral springs in the upper reaches of the Baksan and Malka Rivers (Vasin et al. 2016).

On the northern slope of the Elbrus, at an altitude of 2,380 m a.s.l., on the bank of the Malka River, there are exits of warm Narzan, or Dzhily-Su (Fig. 3.8), described by G. Abikh (1849), A.P. Gerasimov (1908) and E.E. Karstens (1905) (Vasin et al. 2016). Water of Dzhily-Su springs is carbonic hydrogencarbonate -chloride calcium-sodium type, with a high iron content exceeding the specific norm, with carbon dioxide content of 1.6 g/l, with mineralization of 3.1 g/l, and temperature of 22.4 °C.

There has been a significant amount of interest in the carbonated mineral springs of the Narzan Valley or Khasaut Narzans located 40 km from Elbrus, where 17 mineral springs are located within 3 km. This is carbonic water with hydrogencarbonate - chloride sodium-calcium-magnesium chemistry, with mineralization of 2.3–8.3 g/l. At a distance of 5 km from Narzan Valley, in a valley called Solnechnaya because of its abundance of sunlight, there are many carbonic hydrogencarbonate -chloride sodium-calcium-magnesium mineral springs with mineralization of 1.0–2.2 g/l, commonly known as the Bolshe-Lakhransky springs. In Kabardino-Balkaria, there is much Borjomi-type carbonic hydrogencarbonate -sodium mineral water and Essentuki-type carbonic hydrogencarbonate -chloride-sodium water. There are many hydrogen-sulfide springs represented predominantly by cold sulfate–calcium water with low hydrogen sulfide content. This type includes the Belorechensk springs near Nalchik, which are widely used in balneological practice (Vasin et al. 2016).

Fig. 3.8 Kabardino-Balkaria, spring of Dzhily-Su mineral water

3.1.6 Republic of North Ossetia (Alania)

This republic is unique in its reserves and the types of mineral water occurring here. Currently, there are about 300 springs and deposits of carbonic hydrogencarbonate -chloride sodium boron water. According to their balneological properties, chemical and gas compositions, the mineral springs of North Ossetia have such widely known analogues as Matsesta, Essentuki, Borjomi, Truskavets, Kislovodsk springs, etc. (Misetov and Polkvoy 2012; Bolatati 2013).

In Republic of North Ossetia (Alania), 20 mineral water deposits are currently known. There are confirmed reserves for the following deposits: Nizhne-Karmadon, Tib, Tamis, Zaramag, Korinsky, Zamankul, Redant, Razdolnoe, Mamison, and Biragzang and FAT sites (Misetov and Polkvoy 2012; Bolatati 2013). A large number of mineral springs, diverse in their chemistry and properties, provide the objective prerequisites for the development of a resort and health economic sector in the republic, as well as restoration of existing health facilities and construction of new complexes. In addition, a base exists for a substantial increase in the volume of industrial bottling of table and medicinal-table mineral water in the region.

3.1.7 Republic of Ingushetia

Despite its small area, the Republic of Ingushetia has significant reserves of mineral water, and many springs have been used by its locals for many centuries.

Silver spring (Khyovr-Khiy). The exact meaning of "*khyovra*" is not known, but the approximate meaning is "healing." This is a legendary spring located between the ancient villages of Bisht and Kyazi. The Khyovr-Khiy spring of life-giving water (Fig. 3.9). Locals used the water to treat many diseases, fractures, and injuries, and as a remedy for heartburn. There are also cases known of healing from severe diseases of the kidneys, liver, bladder, heart and blood vessels, as well as cancer. In the 1980s, St. Petersburg scientists conducted an examination of the Khyovr-Khiy water source and found that it contains silver. It is well known that silver is a powerful disinfectant, suitable both for the treatment of external injuries and for internal use; it cleanses the body of harmful microbes. The famous Russian doctor N.I. Pirogov, who studied the Caucasian hillfolk's methods of treating gunshot and purulent wounds, wrote about using silver water in the Caucasus for the treatment of fistulas and ulcers with decay and suppuration resulting from bone tuberculosis of the lymph glands.

Obankhi spring is located in the famous Obansky Gorge and emerges at an altitude of 3,000 m a.s.l., from the specially protected mountainous Ingushetia historical-architectural natural reserve-museum (Fig. 3.10). One liter of Obankhi water contains more than 20 cations and 15 anions included in the periodic table. There are even ultra-microelements in Obankhi water such as silver and gold, which have a bactericidal effect on the body. The unique properties of the water from this spring are confirmed by environmental experts from Moscow and Pyatigorsk. The water is recognized as one of the purest in the Caucasus. It does not form scale when boiled.

Fig. 3.9 Silver spring (Khyovr-Khiy)

Fig. 3.10 Obanhi spring (Eban village)

The Achaluk deposit is located on the northern slopes of the Sunzha ridge and has been known since the Middle Ages. Russian chronicles mention that Achaluk is a town where the "bitter river" flows and the headquarters of the Uzbek Khan were located. Throughout the Alkhanchurt Valley, in gullies, gorges, and ravines, many springs of mineral water and mud are scattered. Particular attention is drawn to the hot springs that are called "mel khiy." Passersby often stop here to stock up on its healing water. In the Achaluki No. 2 spring, there is mineral water of the hydrogencarbonate -sodium type, with mineralization of 2.8–3.3 g/l; it is classified as medical-table and is used for upper respiratory diseases and gastrointestinal disorders.

3.1.8 Chechen Republic

In terms of the number and variety of mineral springs, as well as sites for the placement of landscape and climatic resorts, Chechen Republic occupies a leading position in Russia. The healing power of the local mineral water has been known for a long time. Long before Russian researchers officially discovered the "hot greenhouses" in Chechnya, numerous thermal mineral springs were already being used by locals for treatment. In the hillfolk's tradition, the water endowed with magical properties has been respected since early times (Vasin et al. 2017; Daukaev et al. 2017).

The mineral springs between the Terek and Sunzha Rivers were also used by the Cossacks, who settled on the left bank of the Terek River in the sixteenth century. Early sources include a report from 1586 that a group of Cossacks was located at the Goryachiy Kolodets on the Tersky ridge (in the area of the modern village of Goryacheistochnenskaya). In the documents of the seventeenth century, it is mentioned that locals use this healing water.

Hot mineral springs known as "Terek greenhouses" were studied and briefly described by the physician in ordinary, doctor of medicine G. Schober in 1717 on the personal order of Peter the Great. He named them the "St. Peter's springs" and recognized that their healing properties were applicable for the treatment of many diseases. It is believed that Peter the Great himself visited these springs in 1722 when he returned from the Persian campaign. Officially, these springs were discovered earlier than the hot mineral springs at Mount Mashuk in Pyatigorsk (1780) (Romanova and Kerimov 2017).

By the beginning of the twentieth century, significant resources of mineral water with various physicochemical properties were identified in Chechen Republic (Vasin et al. 2017; Daukaev et al. 2017). Well-known mineral springs include: Grozny, Sernovodsk, Gudermes, Bragun, Argun, Vedeno, Melchkhi, and others (Fig. 3.11).

The well-known Sernovodsk-Kavkazsky resort deserves special attention. In the Caucasus, after the resorts of the Caucasian Mineral Waters were designated as government facilities (1803), the next to be established there was the Mikhailovsky (Sernovodsky) resort belonging to the Caucasian Linear Cossack Army, and after 1860 it belonged to the Terek Cossack army. In 1845 and 1846, the Cossack village of Mikhailovskaya (the current Sernovodsk) was founded on military lands, and the

Fig. 3.11 Natural hydrogen sulfide spring in Melchha gully. Chechen Republic, Gudermes district

springs were named the Mikhailovsky, and later Sernovodsky, hot mineral springs. Here, in almost the same territory, four types of mineral water were discovered: sulfide, hydrogencarbonate -sulfate-sodium, chloride-sodium, and iodine-bromine (Vasin et al. 2017).

3.1.9 Republic of Dagestan

Dagestan is rich in mineral water springs that vary in temperature, chemical and gas composition. According to the data of V.V. Shtilmark and O.A. Bairashevsky, more than 67 mineral springs were registered in the republic before 1947. As a result of studies conducted by the geochemical laboratory of the Institute of Geology of the Dagestan Scientific Center of the Russian Academy of Sciences and others, the number of mineral springs described reached 286 (Alieva 2017). Among them are such healing water types as sulfate-hydrogencarbonate -sodium thermal water containing trace elements (iodine, bromine, fluorine, potassium, iron, boric, silicic acid), iodine-bromine highly mineralized water, iodine-bromide chloride-sodium water, thermal water with a high content of trace elements, and hydrogen sulfide water.

Since the nineteenth century, water has been used for medicinal purposes in Dagestan from Nizhnee Inkhelo and Kvankhidatl springs in the Botlikh region, as well as the water of the Tindin and Inhokvartin carbonic boron springs in Tsumadin district; Khzanor spring in Tsuntin district; Agvali, Bezhta, and Tidib soda water; Chokh and Karatakh ferruginous water; Khvatikunin, Tlyaratin, and Gergebil springs, similar in salt composition to Kislovodsk Narzans; and Korodin and Datun healing springs in the Soviet district. The best known of them are the healing springs near Petrovsk, Khasavyurt village; sulfuric spring in the Eki-Bulak mountain area near Temir-Khan-Shura; and especially the Kayakent springs and mud (Alieva 2017).

The main types of mineral water in Dagestan are hydrogen sulfide, hydrogencarbonate -sulfate-sodium, hydrogencarbonate -chloride-sodium, and chloride-sulfate-sodium (Alieva 2017). The hydrochemistry of the underground mineral water is largely determined by the structural and tectonic zonality of Dagestan. So, in the coastal part of Dagestan, there is widespread highly mineralized chloride-sodium water with a high content of trace elements iodine, bromine, and boron. In the mountainous part, low mineralized water with a low content of trace elements is common. In central mountainous Dagestan, highly mineralized water is adjacent to weakly mineralized sulfate–chloride-sodium-calcium water.

3.2 Crimea

Mineral and fresh underground water. Crimea has a significant hydromineral resource potential, including reserves of medicinal mineral water, brine, and medicinal mud. In total, about 150 water sources of various types have been recorded on the Crimean Peninsula. Mineral water located in the region is underused in the Crimean resort industry. Currently, operational reserves have been confirmed at five deposits (Saksko-Yevpatoria thermal water, Feodosia water, Chokrak hydrogen sulfide water, Yevpatoria sea-type water, and Novoselov thermal water). The total reserve of mineral water is 29,299 m^3/day. In addition, five deposits are being operated without confirmation of reserves (Yalta, Aivazovskoye, Mellassa, Almazny resort, and Adzhi-Su). Their resources are estimated at 1,112 m^3/day. (Yakovenko 2015; Mineral groundwater 2018).

Crimean mineral water is diverse in origin, depth, reserve, flow rate, chemical and gas composition, and temperature; it is distributed in aquifers from the Quaternary to Paleozoic ages and has a wide geography (Map 3.2).

Water without specific components is most widespread on the peninsula in all its structural and tectonic regions. By mineralization level and chemistry, the water of this balneological group in Crimea can be divided into four types:

Nitrogen-sulfate water with varied ionic composition, with mineralization level of 0.5–2.0 g/l; it is distributed mainly in foothills and mountains of Crimea. Water seepages are represented by seven springs and three wells.

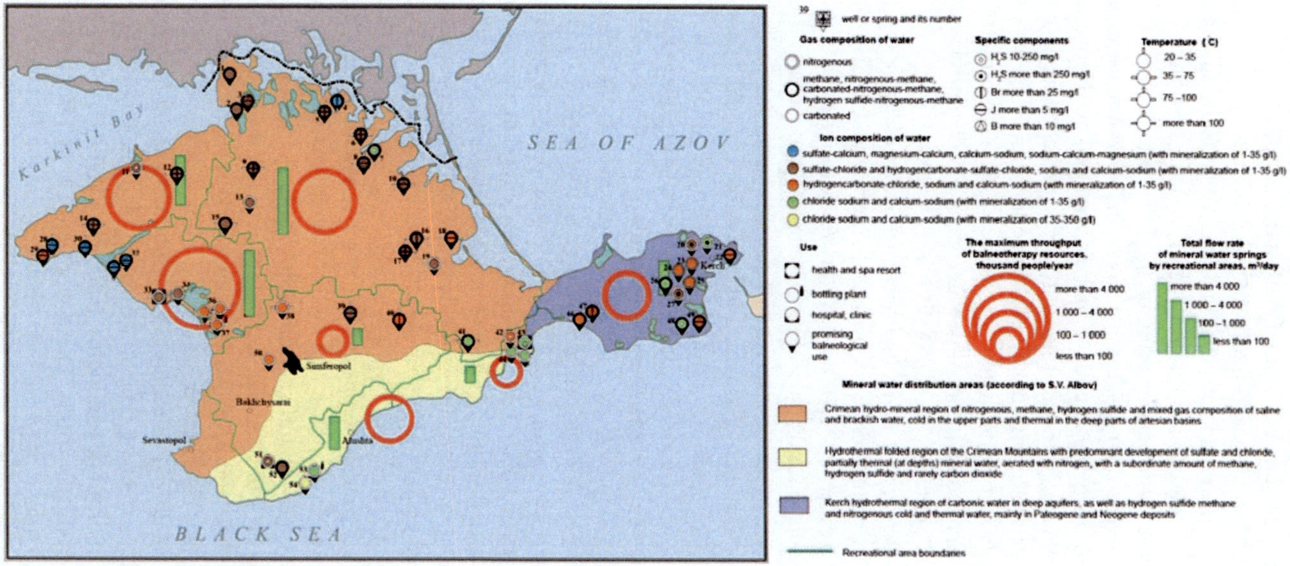

Map 3.2 Balneological resources of Crimea. *Compiled by I.M. Yakovenko*

Within the Belogorsk trough, the flow rate of these springs is quite insignificant (up to 0.1 l/s). In the mountainous part of Crimea, mineral water of this type is found within the Sudak synclinorium and South Coast anticlinorium.

The freshwater uncovered during the boring of the Yalta Tunnel and used until recently for bottling by the Yalta Brewery is well known. In addition, there are rather sufficient springs of freshwater such as Pripigadi, Ai-Yori, and Savlukh-Su on the southern coast of Crimea, and Laura spring on the Belogorsk slopes.

Izhevsk-type and Feodosiya-type *nitrogenous chloride-sulfate water* with mineralization of 2–10 g/l is quite abundant and is confined to Neogene deposits. Within Crimean Mountains, mineral water with described chemical composition is distributed in the eastern part, within the Sudak synclinorium. The Feodosia deposit, represented by springs and wells, is the best known. Since the mid-1970s, a deterioration of the sanitary indicators of this water has been observed and its use for medicinal purposes is limited.

Nitrogenous, nitrogenous-methane, hydrogencarbonate -chloride sodium thermal water with mineralization of 2–5 g/l is very widely represented in Crimean lowlands and is found in deep wells (500–1,400 m). At the Yevpatoriya deposit, water of this type is found in six wells in sandstones of the Albian Stage of the Lower Cretaceous, and is used by a number of resorts as a medicinal-table water.

The Saki deposit, confined to sands and sandstones of the Hauteriv-Barrem (Neocomian) Lower Cretaceous, is found in four thermal water wells at depths of 900–1,001 m. All four wells are widely used for the production of mineral water, both for medicinal purposes and industrial bottling. Water of this deposit, called Crimean and Crimean Crystal is widely known in Russia in plastic and glass containers.

Mirgorod-type *nitrogenous and methane-nitrogenous chloride-sodium water* with mineralization of 2–10 g/l is widespread on Crimean steppe. It is contained in sediments of different ages and lithological compositions within the Novoselovsk uplift. In Yevpatoriya, this water is found in Paleozoic limestones and Alba sandstones and is used by many children's resorts for medical purposes. The area between the village of Nikolaevka and Saki is promising for obtaining water of this type; three wells are found there.

Thermal water of Crimean steppe, having various chemical and balneological properties, is found in Novoselovsk and Oktyabrsk regions in 25 deep wells (from 1,200 to 2,200 m); it has a significant daily flow rate, but it is practically unused for its intended purpose. For example, the flow rate of the well 38 OK (Pyatikhatka village) is 4,060 m^3/day, well 22 N (Vodopoynoye village) −3,020 m^3/day, and well 20 N (Trudovoye village) −2,160 m^3/day.

Staraya Russa-type *nitrogenous chloride-sodium water with high mineralization* of 10–35 g/l, as well as Usolsky-type underground methane chloride-sodium brines with a high content of bromine and iodine and mineralization of more than 35 g/l, is found in various regions of Crimea. These types have a well-defined balneological purpose, but have not found practical therapeutic use so far. Detailed studies of this water have not been carried out.

Carbonated water. This mineral water group is found only on the Kerch Peninsula. The Gornostaevsky and Tarkhansky sites are the best known, represented by springs located northeast of Kerch. The water of these springs has a very complex and variable chemistry and elevated content of therapeutically active elements bromine, iodine, sulfites, and free carbon dioxide. The flow rate of these springs is low. Their water has been used for self-medication for centuries, but it has not received modern balneological assessment and application.

Sulfurous water. Water of this type is also confined to the Kerch Peninsula and is mainly associated with the Chokrak deposits. Sulfide water is represented at 29 water points (14 wells).

Matsesta-type very strong chloride-sodium sulfide water with mineralization of up to 35 g/l is especially widely represented in the Chokrak deposit. The total reserve of this deposit is 380 m^3/day. The Chokrak resort was successfully operated on the basis of the sulfide water and treatment mud of this deposit until 1936. Nearby, the Karalar and Syyurtash springs of sulfide water are located. Despite their obvious and established balneological value, these water springs have no practical application.

Therapeutic mud (peloids). So far, the Crimean Peninsula has maintained its lead in the amount of first-class highly mineralized sulfide-type therapeutic mud in its salt lakes. Silt deposits of coastal lakes on the peninsula (products of modern geological processes) began to form about 7,000 years ago. Processes of mud formation continue in the lakes to this day, mainly due to the replenishment of microorganisms in the brine and the silt itself with organic matter.

In six lakes of Crimea (Uzunlar −9.1 million m^3, Sakskoye −4.8 million m^3, Chokrak −4.4 million m^3, Tobekik −4.4 million m^3, Dzharylgach −3.0 million m^3, and Koyash −1.7 million m^3), 28 million m^3 of treatment silt is concentrated; in Eastern Crimea, there is 20.2 million m^3; and in Western Crimea, there is 7.8 million m^3. Only the therapeutic mud of the Saki deposit has received a modern balneological certificate from the Ministry of Health and is extracted on the basis of a full set of legal documents for subsoil and water use. Since the canceling of the license for its subsurface use in 2004, Chokrak Lake therapeutic mud has been plundered on a large scale by underground methods. Other deposits went unclaimed for several decades, as a result of which anthropogenic pollution and gradual uncontrolled degradation of the resources occur intensively there.

In the North Crimean group, two of the seven lakes (Kerleut and Aigul) have not yet been destroyed by the chemical industry. Both lakes are promising for recreational and health-improvement purposes, but they are practically unstudied so far.

Salt lakes. The practical use of hydro-mineral resources of coastal salt lakes is associated with the historical development of human society (Fig. 3.12). In the early and subsequent stages, they were, first and foremost, a vital source of edible salt, but they eventually began to be used for the healing properties of their bottom sediments and highly saline water (brine). It has been proven that the highly mineralized sulfide silts of salt lakes have a beneficial effect on the body and health of animals and humans. The scientific study of the medical characteristics of Saki Salt Lake was begun by Professor Desser in 1807. He was the first to perform a chemical assay of

Fig. 3.12 Saki. Museum of the History of Mud Treatment

the silt sulfide mud, and Dr. S.N. Auger continued his studies and obtained positive results for their use in medical practice in 1826 and 1827 (Pasynkov et al. 2014; Ivashchenko et al. 2018). In 1826, a pavilion for the treatment of patients began to operate, and a branch of the Simferopol Military Hospital was opened in 1837. Saki mud was popularized for the treatment of rheumatic diseases, tuberculosis of the bone, other diseases of the bones and joints, chronic syphilis, and infertility (Fig. 3.13). In 1828, the Saki resort received 29 officially registered patients, but there were ten times more towards the end of the nineteenth century (Bezrukov 1998; Yakovenko 2015).

The biotechnological, pharmaceutical, and food industries of the world are extremely interested in obtaining biologically active substances (carotenoids, amino acids, vitamins, polyunsaturated fatty acids, etc.). These natural substances are found in the microalga *Dunaliella salina* and small crustacean *Artemia salina*. These organisms, which purify brine and contribute to mud formation, are extremely prolific, and they completely dominate in the brine in the summer-autumn season, creating colonies constituting hundreds of tons of first-class microbiological raw material. Meanwhile, this wealth of the Crimean lakes is used only in the most primitive way: for feeding aquarium fish.

Global production of carotene is about 1,430 tons per year, but only 35 tons of natural (more effective than synthetic) beta-carotene is produced. World market demand is only satisfied by 2.5–3.0%. The minimum price for 1 kg of crystalline

Fig. 3.13 Saki resort. Vintage baths for mud treatment

beta-carotene is $1,800, and especially pure product is worth up to $20,000. Among all carotene carriers, the most productive source of natural beta-carotene and vitamin A is acknowledged to be the algae *Dunaliella salina* living in salt lakes.

In China, Australia, and the United States, technologies for the intensive cultivation of this algae and industrial production of natural beta-carotene from it have been developed. The steady growth of demand for this product is forecasted on the world market.

According to experts, the volume of natural carotene production in the Crimean lakes and Sivash could be 120–150 tons per year, or 10% of the world market, with the possibility of increasing production to 800–1,000 tons per year.

No less valuable is another representative of saline biota, the crustacean *Artemia salina*. Currently, there are patents for a number of water extracts, biologically active additives, drugs, and emulsions obtained by processing both the crustacean and its reproductive products (cysts and nauplii) for the production of antiviral, anti-herpes, vulnerary, fortifying, anti-cancer, and other drugs and cosmetics.

The use of the microbiological resources of salt lakes seems to be a very promising type of economic activity, guaranteeing access to the world market and producing large profits.

Mud volcanic breccia (pelites). On the Kerch Peninsula, among 28 small manifestations of mud volcanism, there is a deposit of pelites with a high content of boron, iodine, and bromine: the Bulganak volcanic field. Rocks liquefied and frayed by deep tectonics have been rising to the Earth's surface from depths of 7–8 km and accumulating in the folds of the Kerch Peninsula for hundreds of thousands of years. The Bulganak volcanic field is a boron province containing significant concentrations of many therapeutically active trace elements in the chemistry of its extremely finely divided pelites.

Bulganak pelites have been used for treatment since antiquity. From the end of the nineteenth century to the Great Patriotic War, the Chokrak-Bulganak institution in Kerch for mud cures operated successfully using this mud volcanic breccia. At the same time, boron and iodine were extracted industrially from two major volcanoes, the Oldenburg and Central. At present, no economic activity is being conducted at the ownerless and unprotected deposit.

In 1976, the Geominvody expedition determined that the renewable reserves of the deposit amount to 5,000 m^3 per year. The Saki GGRES tentatively determined reserves of compacted balneologically valuable pelites to be up to 1.5 million m^3.

Bentonite clay (keel). Among varieties of medicinal clay, bentonite clay (the Crimean name is keel, kefekelit) is considered an elite product. Keel has been used in Crimea since ancient times for both domestic and household purposes, as well as in traditional folk medicine.

Crimean bentonite clay is formed as a result of an interaction of ancient ocean water with volcanic rocks about 500 million years ago. The keel lies among Upper Cretaceous marls in the form of a layer of the Second Ridge of Crimean Mountains, from Sevastopol to Simferopol and Belogorsk. However, this layer rarely comes close to the surface. Intensive extraction of the keel for export took place from the sixteenth century to the beginning of the twentieth century, depleted its reserves: the Sapun-gora and Kurts deposits have been fully developed. Today, only the Kudrinskoye deposit has survived, with an estimated reserve of 600,000 m^3. Bentonite formations are likelt to be detected in the vicinity of Prolom village (Belogorsky district) and in the Kerch region.

Salt lakes have long been the main raw material base for the galurgic industry, which produces a complex of various bromine-containing and magnesian products, as well as the world's best edible salt containing biologically active compounds, which makes it possible to recommend it for widespread use in medical practice to obtain a healing effect comparable to the use of pharmaceuticals.

As modern biochemical studies have shown, the therapeutic properties of brine are almost completely preserved in the salt obtained from it. In other words, brine (sea) salt is a long-term therapeutically active concentrate of live brine.

Deep under Crimea, a large volume of deep salt (sometimes thermal) brines is contained, a detailed study of which has not yet been carried out, but, by analogy with foreign experience, there are undoubted prospects for its therapeutic use.

Lake and deep brines. Sea (brine) salt. The natural brine of Crimea salt lakes is the repeatedly concentrated water of the Black Sea, with mineralization of up to 250 g/l, enriched with salts from ascending mineral springs. In the millennia-long history of seasonal evaporations, therapeutically active trace elements have accumulated in the lake brine in concentrations exceeding their content in seawater by seven to ten times. Lake brine contains cations (sodium, potassium, magnesium, calcium), anions (sulfates, chlorides, hydrogencarbonates), and trace elements (bromine, iodine, phosphorus, arsenic, boron, nickel, silver), as well as biogenic stimulants (amino acids, hormones, vitamins, lipids).

Coastal lakes. A peculiar feature of Crimea is a coastal lake-lagoon-estuary strip, which includes about 300 saline reservoirs with specific chemistry of their water and bottom sediments. Forty-eight large salt lakes are known in the region, 26 of which have an area of more than 1 km^2. The properties of the water and bottom sediments in these reservoirs differ depending on location, water regime, freshwater inflow through small rivers and gullies, presence or absence of connection with the sea, and their chemistry.

In 2017, employees of the State Unitary Enterprise of Republic of Crimea "Crimean Hydrogeological Operational Station" prepared the Cadaster of Deposit of Coastal Lakes of Republic of Crimea, informational reference material on their current status for their effective economic use (Vasenko et al. 2017). Currently, four main groups are territorially distinguished among the coastal lakes of Crimea: coastal lakes of the Perekop group, lakes and bays of the Tarkhankut group, lakes of the west coast, and coastal lakes of the Kerch group. Among them, the natural resources of five Crimean deposits have proven healing properties: the Saki Lake deposit of therapeutic mud and brine, the Chokrak Lake deposit of therapeutic mud and brine, the Koyash Lake deposit of therapeutic mud and brine, the Dzharlylgach Lake deposit of therapeutic mud and brine, and the therapeutic mud volcanic breccia and pelites of the Bulganak mud volcanic manifestation.

3.3 Tyva

Tyva Republic has a unique geographical location, at the junction of South Siberian taiga forests and Central Asian steppes and deserts. The features of its relief, climate, geological structure, and hydrogeological conditions predetermined the distribution of various types of natural medicinal water in the republic: arzhans (healing springs), and saline and mud lakes. In Tyva, almost all types of mineral water known in the Russian Federation and the former Soviet Union can be found (Ivanov 1968).

The hydromineral resources of Tyva can be divided into six separate hydromineral regions (Map 3.3). In these regions, types of mineral water such as nitrogenous terms, carbonic cold and thermal, hydrogen sulfide cold and thermal, radon, saline, acidic, siliceous, arsenic water, etc., can be distinguished.

The study of Tyva's hydromineral resources was carried out by many scientists, in particular by V.M. Levchenko (during the 1930s), A.O. Blumenfeld (1950s), E.V. Pinneker (1960s), and others. These studies were predominately devoted to the macrochemical parameters of Tyvan mineral water.

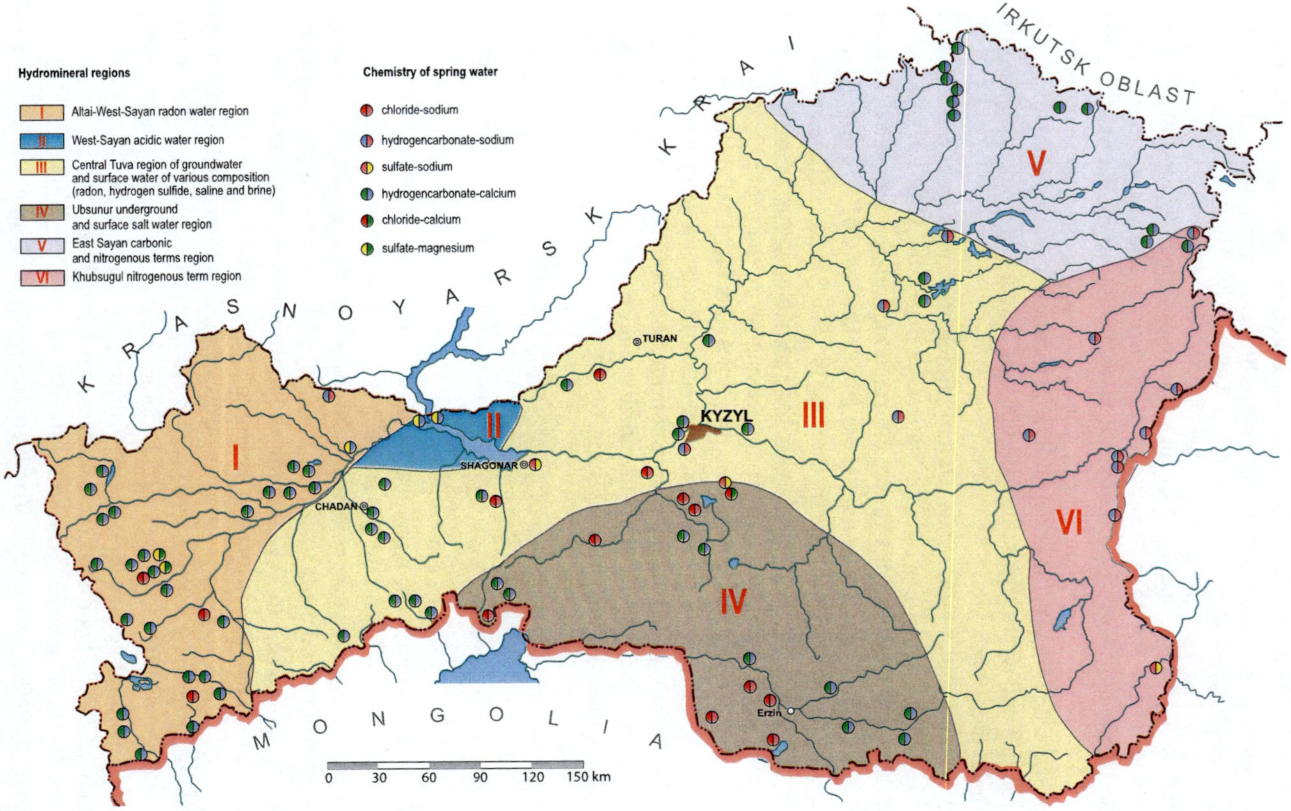

Map 3.3 Springs of mineral water in Tyva. *Compiled according to the map "Springs of Tyva Republic" provided by the Ministry of Health of Tyva Republic's Research Institute of Medical and Social Problems and Management of Tyva Republic*

Since the late 1980s, Tyvan scientists (K.D. Arakchaa and K.S. Kuzhuget), along with scientists from Irkutsk State University (G.M. Speiser and others) and Tomsk Polytechnic University (M.B. Bukaty and others), have been conducting systematic, comprehensive studies of the hydromineral resources of Tyva. Research has covered almost the entire area of the republic. Hydrochemical, geophysical, microbiological, ecological, and heliometric studies have been carried out. In total, more than 70 mineral and fresh arzhans were studied for 50 different parameters.

Every year, up to several tens of thousands of people are spontaneously treated in Tyvan arzhans (Fig. 3.14). This phenomenon of arzhan treatment, which has been an integral part of Tyvan culture and traditions since ancient times, is a unique phenomenon.

Tyvan natural medicinal water resources can be divided into two large groups:

- *mineral medicinal water*, which includes salt and mud lakes and arzhans falling under the definition of natural medicinal mineral water according to their mineralization level, ion-salt chemistry, and the presence of biologically active components (mineral arzhans);
- *fresh medicinal water*, which includes arzhans not falling under the definition of natural medicinal mineral water by their mineralization level and because of an absence of biologically active components, but widely used for medicinal purposes by the locals through the ages (fresh arzhans).

At this point, more than 200 natural water treatment facilities have been registered in the republic: 74 with medicinal mineral water, including 54 groups of mineral arzhans, as well as 128 fresh arzhans, and 20 salt and mud lakes.

The therapeutic effect of salt and mud lakes and mineral arzhans is mainly due to three types effects on the human body: mechanical, temperature, and chemical.

The mechanical effect is the water pressure on a body immersed in a bath, i.e., hydrostatic pressure. It affects the microcirculation system, increasing transcapillary

a b

Fig. 3.14 Tyvan arzhans—Iyispen arzhan (**a**), Dashtyg arzhan (**b**)

(intercapillary) metabolism in the blood and lymph circulation systems, activating interactions between capillary systems and extracellular fluid, and stimulating an exchange between intracellular and extracellular space.

Thermal exposure contributes to the expansion of blood vessels in the skin, causing significant blood redistribution in the body. Depending on the bath temperature, the functional state of the central nervous system also changes. Warm baths lower its irritability, aiding in falling asleep; short-term hot baths act excitingly, increasing emotional tone. Activation of the metabolism and redox processes in tissues only occurs under the influence of hot and cold procedures. Warm procedures do not have as strong an effect.

The chemical effect is one of the significant balneofactors. Natural mineral water and therapeutic mud containing mineral and organic components in increased amounts, actively influence the functional state of the human body. Balneofactors can be ion-salt composition, and the presence of biologically active components (organic matter, radon, silicic acid, boric acid, bromine, iodine, iron, arsenic ions, etc.) in necessary amounts.

To identify the possible therapeutic effect of Tyvan arzhans, a comprehensive study of Tyvan natural water healing resources and the arzhan treatment phenomenon has been underway since 2010. The study is being carried out by the Laboratory of Arzhanology and Tourism (ArzhanLab LLC) and the Scientific Research Institute of Medical and Social Problems and Management of Tyva Republic (Research Institute of MSPM RT).

Hydrogeochemical studies of natural water treatment facilities made it possible to distinguish three groups of Tyvan arzhans: mineral salt-mud lakes (20), mineral arzhans (52 deposit groups) and fresh arzhans (more than 130 springs).

Salt-mud lakes have mineralization of 1.5–300 mg/l; according to their main ion-salt composition, they are sodium chloride, chloride-sulfate magnesium-sodium, hydrogencarbonate-chloride sodium, etc. (Fig. 3.15).

In the group of mineral arzhans (mineralization more than 1 g/l), alkaline soda (pH 9.8–10.0), acidic (pH 2.6–3.6), radon (100–1300 Bq/l), carbon dioxide (500 and more mg/l), and hydrogen sulfide (more than 50 mg/l) springs are distinguished, as well as mineral springs without specific properties and components. All these arzhans can be classified as medicinal-table and medicinal water.

In the group of *fresh arzhans* (mineralization less than 1 g/l, without specific properties and components), water of priority drinking quality, of the highest and first quality category, was distinguished. All fresh arzhans can be classified as natural table water.

Biomedical studies have shown that, in treatment with fresh arzhans, as well as mineral arzhans, a positive effect is observed for many diseases. This fact makes it possible to attribute fresh arzhans, which in the mid-twentieth century were characterized by some scholars as false arzhans, to Tyvan natural healing water.

A study of the radiological situation in the Tyva arzhan regions showed that the radioactivity index is within the normal range: from 10 to 40 μR/hour. The possibility of a positive effect from the weak radioactivity of natural water on the human body, adequate for the radioactive balance of the human body, is noted.

Fig. 3.15 Cheder Lake has been used by Tyvan locals for medicinal purposes since ancient times

3.4 Transbaikal Region

The Transbaikal region is one of the largest regions of global significance in its reserves and potential resources of mineral water; it occupies one of the leading places in Russia, second only to the Caucasus. Here there are almost all types of mineral water, from thermal nitrogenous to cold carbonic and from low-mineralized to brackish, as well as hydrogen-sulfide radioactive water. Researcher K.N. Tulchinsky wrote that the exploitation of these springs as resorts would make Transbaikal not only an all-Siberian health resort, but Transbaikal could fill all of Siberia, from the Urals to the Pacific Ocean, by exporting mineral water (Tulchinsky 1919).

The native population of Transbaikal has long used mineral water for various diseases, transferring the experience of their use from generation to generation. Russian Cossacks who settled in Eastern Siberia also used the healing power of the mineral and thermal springs.

In the Transbaikal region, there are about 300 springs of cold carbonic and thermal nitrogenous water. Currently, 20 deposits of mineral water have been explored and evaluated.

The largest deposit is Darasun, with an operational reserve of 520 m^3/day and a water withdrawal of 280–330 m^3/day.

Health and spa resorts operate at nine deposits (Darasun, Molokovka, Kuka, Urguchan, Sivanda, Yamarovka, Yamkun, Karpovka, Zymka-Arshan, etc.).

According to its chemistry and temperature, the mineral water is divided into: (1) cold carbonic, (2) cold fresh radon, (3) fresh ferruginous, (4) cold fresh hydrogen-sulfide, (5) thermal nitrogenous, and 6) thermal nitrogenous-carbonic. The distribution of mineral water springs of the Transbaikal region is shown on the map (Map 3.4).

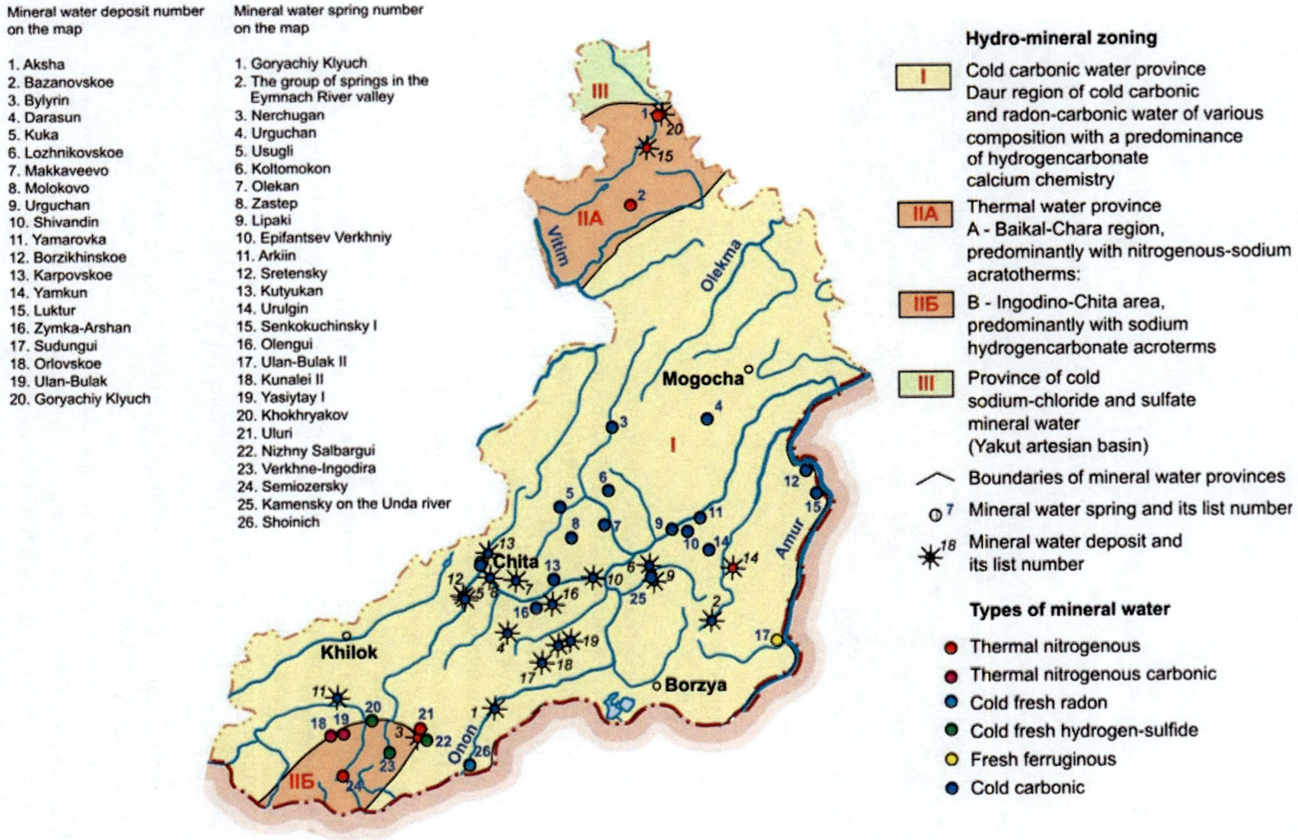

Map 3.4 Springs of mineral water in the Transbaikal region

Carbonated mineral water. As can be seen on the map, the most common is cold carbonated mineral water, which makes up 80% of the total number of springs in the region.

Cold highly mineralized water and brines are known only in the far north of the region. Cold fresh radon, ferruginous and hydrogen-sulfide water has more widespread occurrence. There is only one cold hydrogen-sulfide radon spring known – Khokhryakovsky.

Carbonic springs are predominantly very cold, with water temperature 0 to 2–3 °C; springs with a temperature of 8–10 °C are occasionally found, and only Yamkun and Aktaguchi springs have a temperature of 19–20 °C. In gas composition, cold mineral water is predominantly carbonic and mixed (nitrogenous-carbonic or carbonic-nitrogenous). According to its mineralization level, carbonic mineral water is divided into fresh (to 1 g/l), low brackish (1–3 g/l) and strongly brackish (more than 3 g/l). Most of the mineral springs have a mineralization of 0.5–0.6 g/l, rarely of 1 g/l. Mineralization of 3 g/l and more has only been established in a few springs (Kuka, Kavykuchin, Zun-Torey, etc.). Carbonic water is divided into several types depending on its chemistry: Darasun (hydrogencarbonate alkaline-earth water), Kuka (hydrogencarbonate sodium-alkaline earth water with a sodium content of 25–50%-eq.), Taseev (hydrogencarbonate -alkaline earth water with a sodium content of 50–75%-eq.), Baley (hydrogencarbonate sodium water), and Torey (chloride-hydrogencarbonate sodium water). In addition, carbonic hydrogencarbonate -sulfate and sulfate water is known with a variable cationic composition.

Darasun-type mineral water is the most widespread in the region. It includes more than 100 carbonic springs. This type of water is characterized by a low mineralization, as well as hydrogencarbonate ion content of up to 80–90%-eq. and alkali amounts less than 25%- eq. Sometimes there is an increased iron content. Sulfate ion is present in small amounts, but prevails over chlorine ion.

Taseev-type hydrogencarbonate carbonic water is close in its chemistry to the mineral water of the Lastochka resort (Primorsky Krai). Water of this type is found in works at Taseevo at depths of 316–366 m, as well as in a well in the Selinda Valley at a depth of 24 m.

Baley-type hydrogencarbonate calcium-sodium and sodium carbonic water is found in the Baley graben at a depth of 200 m and more. Baley-type water includes the Senkokuchin and Arshantuy springs. Water mineralization in these springs is 2–5 g/l, and alkali content reaches 80–100%-eq. Baley-type water is close in its chemistry to the water of the Borjomi spring.

Torey-type carbonic water is found within the Torey Depression filled with Lower Cretaceous terrigenous deposits. In its chemistry, this water belongs to the sodium chloride-hydrogencarbonate type and is similar to the Essentuki-type water of the North Caucasus.

Many carbonic mineral springs are characterized by increased radioactivity, which determines their basic medicinal properties. Cold carbonic water usually contains such microcomponents as molybdenum, lead, copper, aluminum, manganese, titanium, beryllium, and others, in small amounts. In the gas composition of these springs, carbon dioxide exceeds 90% of the total gas volume. In springs with mixed gas composition, methane content reaches 7.7%. In the gas composition of these springs, the oxygen content reaches 5–7%; it is associated with the process of mixing mineral water with ground nitrogen–oxygen water. Rare heavy gases are contained in an amount of 0.01–0.8%, rare light gases are present at 0.01–0.26%. In springs with nitrogen in their gas composition, an increased argon content is noted.

Thermal nitrogenous and nitrogenous-carbonic springs have limited distribution within geographically separated hydromineral areas in the north, southwest, and southeast of the region.

In its chemistry, nitrogenous thermal low-mineralized water is divided into sodium-hydrogencarbonate, chloride-sulfate, sulfate-hydrogencarbonate, chloride-hydrogencarbonate, sulfate–chloride, and hydrogencarbonate calcium-magnesium.

Hot springs in the Chara River valley have long been known to the locals as medicinal. They are confined to a large tectonic fault that borders the Upper Chara Depression on the north. By its chemistry, the mineral water described is sulfate–chloride sodium with a mineralization of about 0.2 g/l. The effluent water temperature is 46–48 °C. Chara springs stand out in a special *Chara type* of Siberian term due to their original water chemistry. In the Baikal-Chara folded region, in addition to Chara-type thermal nitrogenous water, there is thermal highly mineralized water. It includes a group of springs located in the Eymnach River basin. According to its chemistry, the water of these springs is hydrogencarbonate magnesium-sodium and hydrogencarbonate sodium-magnesium. Water mineralization is 1–5 g/l. pH varies from 7.0 to 9.6 g/l. Thermal springs with sulfate-hydrogencarbonate and hydrogencarbonate sodium water are located in the highlands of the Daur and Borshchovochny ridges, as well as in the upper reaches of the Chikoy, Ingoda, Kyra, and Bylir Rivers (Ingodin-Chikoy area). By temperature, Ingodin-Chikoy springs are divided into hot and warm. Only two are hot – Kyrin and Bylyrin.

Warm mineral springs (Semiozersky, Uluri), including 12 wells (Talachin, Esutay, Kunaley, etc.), have a temperature of 17–31 °C. In the thermal water of the Ingodin-Chikoy area, a significant content of silicic acid, as well as the presence of fluorine, rare, and trace elements are noted. By its chemistry, thermal water of the Ingodin-Chikoy area belongs to the *Bylyrin type* of sodium hydrogencarbonate water. A particular feature of the thermal water of this area is its radioactivity. Thermal water with a hydrogencarbonate calcium-magnesium composition is concentrated in the southeastern part of the Daur hydrothermal region, in the Gazimur River basin. It includes the water of the Yamkun, Aktaguchinsk, and Nikolaev springs. The carbon dioxide-nitrogen gas composition and high radioactivity are distinctive features of the thermal water of these springs. Thus, the gas composition of Yamkun water includes 77% nitrogen and rare gases in its total volume, and carbon dioxide amounts to 21.4%.

The geological situation of the warm water outcrop in the southeastern part of the Daur hydromineral region, its gas composition, and radioactivity make it possible to classify this water as the Yamkun type of nitrogenous-carbonic water.

Radioactive hydrogen-sulfide mineral water in the Transbaikal region has a very limited distribution. This type of water is found at the Khokhryakov spring in the Chikoy River basin. The water temperature of Khokhryakov spring is 5–6 °C, and the content of free hydrogen sulfide in it reaches 33 mg/l.

In addition to the main types of mineral water listed in the Transbaikal region, there is *ferruginous mineral and fresh radon water*, which has balneological significance. An example of cold ferruginous water is the Ulan-Bulak (Urulyungyi) spring located in the basin of the Argun River and having a sulfate magnesium-ferruginous composition (Fig. 3.16). Cold fresh radon water is quite widespread in the Daur hydromineral region. Those springs are usually confined to tectonic disturbance areas within the hydrogeological massifs composed of granitoids.

Highly mineralized water and brines are formed in the carbonate and saline sediments of the Siberian platform, within the Chara Bay of the Yakut artesian basin in the far north of the Transbaikal region. They belong to the third province of mineral water. An example of this type of water is the Verkhne-Zhuisky spring, located in the valley of the Zhui River, the left tributary of the Chara River. It is *sulfate–chloride magnesium-sodium water* with mineralization of 60 g/l and a temperature of 0.2 °C.

Fig. 3.16 Ulan-Bulak (Urulyungyi) ferruginous spring. It has long been used by locals for self-treatment and collection of ocher as a mineral paint

3.5 Kamchatka

The Kamchatka Peninsula is characterized by the significant originality of its healing resources of thermal water and medicinal plants, which formed under unique natural conditions. The economic importance of the region's healing resources is increasing every year due to the growth of its attractiveness to domestic and foreign consumers.

The nature of the Kamchatka Peninsula is determined by its location in the Kamchatka-Kuril mountain-folded region, which is part of the Pacific volcanic belt of Cenozoic orogenesis with increased tectonic and volcanic activity. Most of the 150 volcanoes, including all active ones, are located in the eastern part of the Peninsula. In volcanic regions, there are many associated wells and thermal springs, and many of them have aesthetic appeal and are natural landmarks, practically unaffected by human activity.

The locals have long used mineral springs, mainly thermal, to treat many diseases. The healing effect of the thermal water has been established by balneological cerification, as well as assessment of its quality and reserves (Fig. 3.17).

The first descriptions of the healing springs were made by P.S. Krasheninnikov in 1738–1740 and G.V. Steller in 1774. For official medicine, the examination of these mineral water springs began in the middle of the eighteenth century. B.I. Dybovsky (1879) and N.V. Slyunin (1897–1898, expedition of K.I. Bogdanovich) established

Fig. 3.17 Natural thermal water outcrop

the medical approach. Exploration of most deposits and determination of their chemical composition and yield were carried out mainly in the 1930-1960s and in the late 1980s – early 1990s.

In health resort operations combining all types of scientific and practical activities using natural healing resources, including the organization, treatment, and prevention of diseases, thermal and cold springs of mineral water began to be used in the 1850s with the creation of military resorts, including one in the village of Paratunka. There, water is used with a total mineralization of more than 1 g/l (0.1%) or with less mineralization, but with biologically active elements with specific healing properties (iron, bromine, iodine, arsenic, sometimes mercury), combined with high temperature, which enhances the balneological effect (Fig. 3.18).

The features of the distribution of deposits and mineral water yields, their chemistry, temperature, and other characteristics depend on many factors, including the history of the geological structure formation, neotectonic and metamorphic processes, manifestation of young volcanism, as well as ancient and modern climatic conditions. These factors underlie the hydrogeological zoning of the Kamchatka mountain-folded region. First-order hydrogeological structures, artesian basins, and hydrogeological massifs of folded areas are distinguished on the basis of regional geostructure, and their mineral water is distinguished by origin, nutrition, chemistry, mineralization level, temperature, occurrence depth, and other features. All these factors determine the formation of different types of mineral water, their diversity, and their medicinal properties (Map 3.5).

According to this classification, the mineral water of Kamchatka is divided into seven main balneological groups: (1) water without specific components and properties, (2) carbonic water, (3) sulfide water, (4) ferruginous and arsenic water with a high content of metals Mn, Cu, Al, Zn, etc., (5) bromine and iodine water, (6) radon water,

Fig. 3.18 Thermal pool of the Antarius guest house, the village of Paratunka

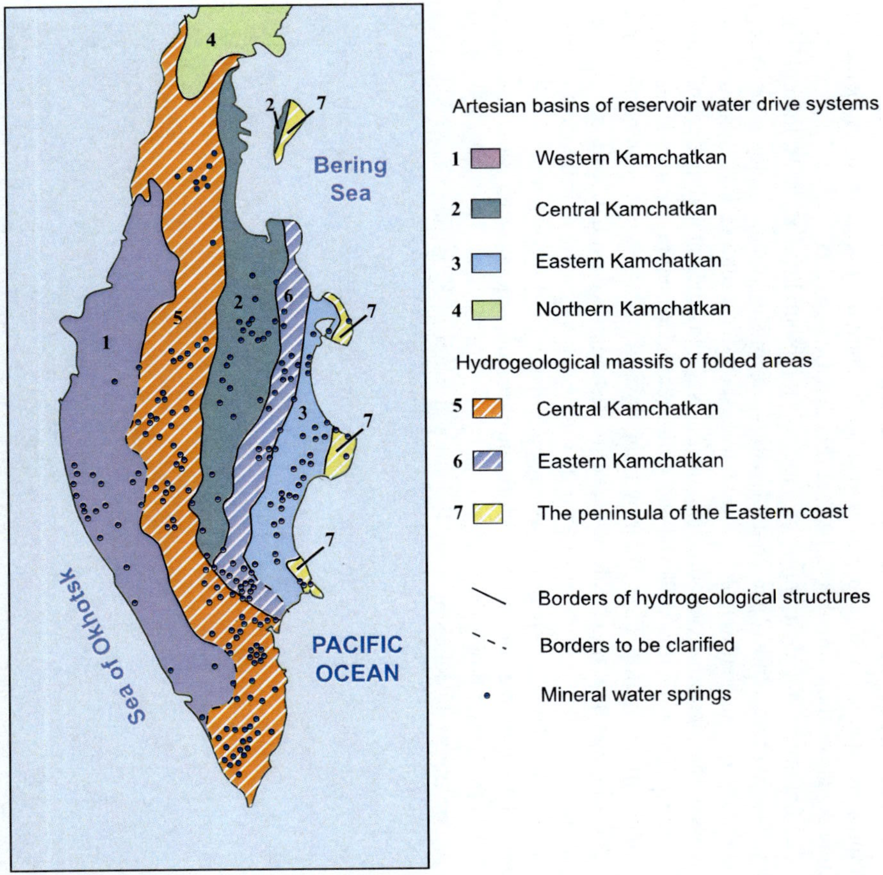

Map 3.5 Springs of mineral water in Kamchatka

and (7) siliceous water. Many types of mineral water here can be simultaneously assigned to several balneological groups (Ivanov and Nevraev 1964).

In Kamchatka, four balneological groups of water are the most widely distributed: without specific components and properties, carbonic, ferruginous, and siliceous (Muradov 2007).

Thermal and cold mineral water without specific components and properties with different mineralization, often bromine and iodine, is developed within artesian pools (1, 2, 3) of the reservoir's waterdrive system.

It has a chloride-hydrogencarbonate sodium or calcium-sodium composition and a total mineralization of 1.2–28.2 g/l, in cold springs it decreases to 0.33–1.17 g/l. The gas composition includes dissolved nitrogen and methane in an insignificant amount (Konmog spring) (Ivanov et al. 1968). The water is used to treat metabolic diseases, diseases of the digestive system, diseases of the genitourinary system, which, respectively, belong to the IV, XI, and XIV categories of diseases according to the international classification.

Carbonated thermal and cold mineral water is chloride-hydrogencarbonate, calcium-sodium, with a total mineralization of up to 5.3 g/l, contains metaboric and silicic acids, iron, and dissolved carbon dioxide (temperature 40–97 °C); it is common in artesian basins of the reservoir's waterdrive systems: Central Kamchatka (2) and Eastern Kamchatka (3), the latter mainly in their southern, volcanic part. Its total potential resource is 27,000 m^3/day. It is used for the treatment of diseases of the digestive system and diseases of the circulatory system (XI, IX categories of diseases). The most famous is the Malka springs (Fig. 3.19).

Thermal sulfide water, with an effluent temperature of 58–100 °C (Verkhne-Semyachik, Ozernovsk, Verkhne-Koshelevs, and others), is low mineralized (0.6–2.5 g/l), strongly acidic, sulfate, often with a high content of H_2S, less often chloride-sulfate, with a large amount of silicic acid (90–350 mg/l), iron (to 195 mg/l), and arsenic (to 2 mg/l) (Ivanov et al. 1968). These types are confined to the southern volcanic part of the Central Kamchatka (5) and East Kamchatka (6) hydrogeological massifs. Their therapeutic profile is balneotherapy; they are used in the treatment of diseases of the circulatory system (IX category).

Ferruginous and arsenic cold and thermal water. Its most famous springs (Nalychevo, Goryachaya Reka, Kraevedchesky, etc.) are located in the southern part of the East Kamchatka hydrogeological massif (6). This water is characterized by a high and very high content of arsenic (5–7 to 20 mg/g) and iron (200 to 400 mg/l); it is chloride-sodium type, containing sulfate and hydrogencarbonate ions in a small amount, and calcium as well. This water is valued in balneology. Its temperature varies from 19 to 75 °C. The main application is in the treatment of diseases of the skin and subcutaneous tissue (XII category).

Bromine and iodine water is typical of the southern part of the Central Kamchatka hydrogeological massif (5). The bromine content reaches 25 mg/l; by its chemistry, this water is divided into carbonic (CO_2 0.74 g/l), bromine (25 mg/l), siliceous (125 mg/l), and hydrogencarbonate -chloride sodium types. The total water mineralization is 6.3 g/l with an effluent temperature of 42 °C; an example of this is Pushchin spring. Bromine water is used for the treatment of diseases of the nervous system (VI category of diseases), and iodine water for atherosclerosis (IX category of diseases).

Fig. 3.19 Malka thermo-mineral springs

Radon water with a radon content of more than 5 mμ C/l (14 Mahe units) encapsulates many balneological groups (Ivanov et al. 1968). The radioactivity of the mineral water of Kamchatka is not well understood. The most famous and commonly used hot low radon springs, with a temperature of 35–45 °C, are also located in the southern, inhabited part of Kamchatka, in the East Kamchatka (3) artesian basin (for example, Nachiki, Paratunka). This water is used to improve metabolism and strengthen the immune system (IV category).

Springs with siliceous thermal water (H_4SiO_4 content more than 50 mg/l, temperature up to 35 °C) are typical of the southern parts of the Central Kamchatka (5) and East Kamchatka (6) hydrogeological massifs of the central and southern parts of the East Kamchatka (3) artesian basin. It is slightly alkaline, chloride-sulfate sodium water with mineralization of less than 2 g/l. These include the Nalychevo, Mutnovo, Malka, and Paratunka springs, as well as the largest deposit in the region – Nachiki springs with a yield of 1,000 m^3/day. By the composition of water-dissolved gases, it is nitrogenous and nitrogenous-carbonic water, less often carbonic and methane. The medical profile of Malka springs is drinking treatment, the profile of the others is balneotherapy. They are used to treat diseases of the musculoskeletal system and connective tissue, diseases of the nervous system, diseases of the skin and subcutaneous tissue, consequences of injury, etc. (XIII, VI, XII, and XIX categories of diseases).

The mineral thermal water outcrop is more often located on the bottoms and slopes of river valleys, in lake basins, and less often on mountain slopes. When rising from great depths, their temperature decreases; lime dissolved in the water in the form of calcium hydrogencarbonate comes in contact with atmospheric oxygen and loses half of its carbon dioxide. Due to its loss, water-insoluble calcium carbonate is released, which is deposited in the form of porous calcareous tuff-travertine, forming relief micro- and mesoforms. Some thermal springs emit silicic acid and form cones or terraces of siliceous tuff. Calcium carbonate and siliceous tuff settle and condense on the walls of the channels along which water rises, isolating them from cold groundwater (Keilgak 1935). Therefore, high-temperature springs can rise from the bottom of a cold river or form channels flowing into "cold" rivers and small rivers, which, for example, reach a width of 15 m at Khodutkino springs, with a depth of 1.5 m. Some deposits of thermal mineral water (for example, Ketchinskoye), which do not have natural outcrops, are accessed by wells.

For regions with active modern tectonic processes, the most characteristic are geysers, the periodically gushing springs already mentioned; for example, the Pauzhet deposit in the southern part of the Peninsula, in the Valley of Geysers, as well as the Uzon deposit in its eastern part. Cone-shaped accumulations of light siliceous tuff (geyserite) are often deposited in the geyser neck with the participation of algae (Ovchinnikov 1963). Many types of thermal springs are regional and federal natural landmarks.

Healing resources also include *mud*—peloids, sedimentary deposits consisting of water, organic substances with thermophilic alkobacterial flora, and minerals. They are influenced by hydrothermal flows along fracture zones and faults, additionally introducing mineral substances that determine their medicinal properties. When heated, they are used in the form of baths and local applications for the treatment of diseases of the nervous, respiratory, digestive systems, diseases of the skin, diseases of the musculoskeletal system, and gynecological diseases (VI, X, XI, XII, XIII, XIV categories of diseases) (Leonova et al. 2018).

The most widely used are sulfide silt peloids common in the most inhabited eastern part of the Kamchatka Peninsula (Utinoe Lake—Paratunka resort), the East Kamchatka (6) hydrogeological massif.

Since detailed surveys of mineral water springs have not been carried out in recent years, a series of modern studies are needed to inventory deposits, study their chemistry and healing properties, and identify new deposits.

While studying and exploiting mineral thermal springs, it is important to take into account their increased sensitivity to external mechanical influences. Outcrops of thermal springs differing in their temperature and chemical chemistry can be adjacent to non-mineralized water, so any careless intervention can ruin their chemistry and reduce flow rate (Fig. 3.20).

Fig. 3.20 Hot tributary of the Anavgay River

References

Alieva PSh (2017) Natural waters of Dagestan and their use in traditional and modern medicine Advancement in of Modern Science 7 200 204 (in Russian)

Bezrukov YuF (1998) Recreational resources and balneology. Simferopol: Simferopol State University, Faculty of Geography, 114p (in Russian)

Bolatati NO (2013) Mineral resources of the Republic of North Ossetia-Alania. In: The development of regions in the 21st century. Materials of I International Scientific Conference. Vladikavkaz: Khetagurov North Ossetian State University, pp 89–92 (in Russian)

Daukaev AA, Gatsaeva LS, Aslan AD, Abumuslimova IA (2017) History of study and use of groundwater and its sources on the territory of the Chechen Republic. Bulletin of the Academy of Sciences of the Chechen Republic 4(37):84–89 (in Russian)

Ivanov VV (ed) (1968) Map of mineral healing waters of the USSR (1:4000000). Moscow (in Russian)

Ivanov VV, Nevraev GA (1964) Classification of underground mineral waters of the USSR. Moscow: Nedra Publishers, 168p (in Russian)

Ivanov VV, Nevraev GA, Fomichev MM (eds) (1968) Map of therapeutic muds of the USSR (1:8000000). Moscow: Ministry of Health of the USSR (in Russian)

Ivashchenko AS, Mizin VI, Ezhov VV, Vasenko VI, Chaban VV, Stankevich DA (2018) Balneological use of therapeutic mud and brine deposits of the western and eastern Crimea in the spa practice. Vestnik fizioterapii I kurortologii, pp 43–51 (in Russian)

Keilgak K (1935) Underground waters. Leningrad-Moscow: ONTI, 494p (in Russian)

Korinevich LA (2015) Mineral waters of Adygea as a basis for recreational activities on the territory of the republic. In: Applied aspects of geology, geophysics and geoecology using modern information technologies. Proceedings of III International Scientific and Practical Conference. Maykop: Kucherenko VO, pp 123–124 (in Russian)

Kotlyakov VM, Badenkov YuP, Galanin AA, Golubchikov YuN, Dunets AN, Kalyuzhnaya YuYu, Kambarova EA, Kruzhalin VI, Mazhar LYu, Mandych AF, Firsova AV, Khropov AG, Chaika IG, Shabalina NV (2017) Where should I go to get to know Russia. Vol 2. Moscow: Codex, 496p (in Russian)

Leonova NB, Miklyaeva IM, Ryabova NV, Malkhazova SM (2018) Actual state of healing resources of the Kamchatka Peninsula and the prospects of their use. Vestnik Moskovskogo Universiteta. Seria 5, Geografia (Moscow University Bulletin. Series 5, Geography) 6:10–17 (in Russian)

Mironov VI, Khodasevich LS (2018) Hydromineral resources of the Krasnodar territory and their medicalapplication. Kurortnaya Meditsina 2:16–24 (in Russian)

Misetov AV, Polkvoy AP (2012) Mineral resources of the Republic of North Ossetia-Alania. Mineral resources of Russia. Econ Manag 1:67–77 (in Russian)

Mineral groundwater (2018) In: State report "On the state and use of mineral resources of the Russian Federation in 2016 and 2017". Moscow: Ministry of Natural Resources and Environment of the Russian Federation, pp 330–331 (in Russian)

Muradov SV (2007) Ecological solution to the problems of modern mud therapy. Petropavlovsk-Kamchatsky: Vitus Bering Kamchatka State University, 266p (in Russian)

Nagalevsky YuYa, Nagalevsky EYu, Zavgorodnaya VP, Tatsenko YuS (2014) Mineral waters as the basis of the balneological complex of Krasnodar Krai Kurortno-Rekreatsionnyi Kompleks v Sisteme Regional'nogo Razvitija: Innovatsionnye Podhody 1 224 227 (in Russian)

Nechiporova TP (2017) Resorts and curative areas of Adygea: current state and development prospects. Kurortno-Rekreatsionnyi Kompleks v Sisteme Regional'nogo Razvitija: Innovatsionnye Podhody 1:208–211 (in Russian)

Ovchinnikov AM (1963) Mineral water (the study of mineral water deposits with the basics of hydrogeochemistry and radiohydrogeochemistry). Moscow: Gosudarstvennoye nauchno-tekhnicheskoye izdatelstvo literatury po geologii I ohrane nedr, 375p (in Russian)

Pasynkov AA, Sotskova LM, Chaban VI (2014) Ecological problems of conservation and use of balneological resources of the salt lakes of Crimea. Uchenye zapiski Tavricheskogo Natsionalogo Universitetay im. VI Vernadskogo. Ser.Geografia 27 (66) 2:97–117 (in Russian)

Romanova OS, Kerimov IA (2017) Terek thermal springs: history of the study and contemporary use. The Herald of the Academy of Sciences of the Chechen Republic 2(35):76–84 (in Russian)

Schaefer GL (ed) (2017) Walking around the North Caucasus. Caucasian Mineral Springs. Travel Guide for pedestrians. Moscow: AST, 128p (in Russian)

Tulchinsky KN (1919) Subsoil resources of Transbaikalia. Chita: Tipografia oblactnogo soyuza zabaikalskyh kooperativov (in Russian)

Vasin VA, Danilov SR, Yeseneev YuI (2015) Resort resources of Karachay-Cherkessia: "gold placer gold" of the North Caucasus. Kurortnaya Meditsina 3:13–18 (in Russian)

Vasin VA, Danilov SR, Chalaya EN (2016) Therapeutic and recreational resources of the Kabardino-Balkarian Republic. Kurortnaya Meditsina 4:17–22 (in Russian)

Vasin VA, Danilov SR, Akhmadov GAN, TZ, (2017) Mineral waters of the Chechen Republic. Therapeutic riches of the North Caucasus. Kurortnaya Meditsina 1:17–22 (in Russian)

Vasenko VI, Gulova OA, Chaban VV (eds.) (2017) Cadaster of coastal lakes of the Republic of Crimea. Types of their economic use (as of 01.01.2017). Saki: KGGRES, 34p (in Russian)

Yakovenko IM (ed.) (2015) Tourist and recreational resources potential of the Republic of Crimea and the city of Sevastopol. Simferopol: IT ARIAL, 408p (in Russian)

Chapter 4
History of Healing Springs in Russia

Abstract The chapter presents historical information about the development of balneology in Russia and the traditions of mineral waters and mud applications. The history of water spring studies can be traced from the time of Peter the Great to the present day. Additionally, the issues of the cartographic study of mineral waters and mud are considered.

Keywords History · Water spring study · Cartographic study

4.1 History of Water Spring Studies

Historically, people have always settled near sources of drinking water, so the search for groundwater, especially far from their natural reservoirs, can be considered as ancient and integral a part of human activity as agriculture. Initially, finding natural water springs that differ in their properties from drinking ones, people gave them magical significance. For example, in the Transbaikal region, many springs are still called "arshans," which means "sacred water" in the Buryat language. In traditional medicine, spring water has been used to treat various diseases through the ages. This is probably the precise reason for the appearance of "living" and "dead" water, which can stop any disease and any patient, in fairy tales, epics, and legends. In world history, there is evidence of the use of thermal mineral water for medicinal and recreational purposes in the Roman Empire, starting in the first century AD (Sergeenko 1960).

In Russia, state development of mineral springs begins with an edict signed by Tsar Peter the Great to the Governing Senate of June 24, 1717, "On the Search for Mineral Water in Russia" (Complete Collection of Laws... 1713–1719).

In 1719, a decree of Peter the Great announced the opening of the first Russian resort—Marcial waters, named after the ancient Roman god of war, Mars. The same decree approved the "Dokhtur Rules," which have become legendary over the past 300 years. These Rules give us the first recommendations on a methodology for using mineral water, observing a regime, proper nutrition, and even walking for therapeutic purposes (Complete collection of laws... 1713–1719).

S. Malkhazova et al., *Healing Springs of Russia*,
https://doi.org/10.1007/978-3-030-83534-7_4

The expeditions to find new mineral water that followed were often associated with the changing military-political situation (military operations in the Caucasus, the Crimean War, development of the Far East). Therefore, military doctors played an important role in the development of national balneology (Starikov 2018).

Thus, inspector of the main military pharmacy G. Schober, on his deployment to the Caucasus in 1719, described for the first time the hot springs in the Terek area, which were used by Russian soldiers. He called them "the greenhouses of St. Peter" and described them as "very healing for many diseases." Subsequently, these springs were called the "Braghun baths." At the same time, Schober became acquainted with the "pretty sour spring" in the region of today's Pyatigorsk and the "sulfur key" in the suburb of Sergievsky in the Astrakhan province. In 1750, on the initiative of the Medical Chancellery, the import of foreign mineral water into Russia was prohibited. In 1798, head doctor A. Levenets and pharmacist I. Guerner submitted a report "On the Properties of Acidic Water" to the Medical Board. They use described water used to treat patients at the St. George Hospital in the Caucasus. As a result, the annual secondment of the doctor was ordered, with his patients sent from the regiments to the warm mineral water of North Caucasus (Shevchenko and Shepin 2003).

In 1800, a book by V.M. Severgin, *A Way to Experience Mineral Water, Compiled from the Latest Observations about This Subject*, was published, in which its author describes mineral springs and lakes in Russia, classifying them according to a combination of attributes and giving detailed recommendations for their research. In 1810, the chief physician of the Department of Home Affairs, P.F. Gaaz, prepared a report to the Minister "Comments on Caucasian Mineral Water," which describes in detail not only the springs, but also the climatic conditions of North Caucasus. He also compiled a list of local plants, adjusting them according to the time of their flowering; this list became an invaluable contribution to the development of national botany. In 1817, a cadaster of the mineral water of Russia was compiled by members of the Mineralogical Society in St. Petersburg under the guidance of Academician V.M. Severgin and Professor D.I. Sokolov.

In 1819, on the order of General A.P. Yermolov, commander-in-chief and head of the civil unit and border affairs in Georgia and Caucasus, large-scale reconstruction and improvement of Caucasian mineral water springs was started, six new bathing rooms were built, including the Yermolov baths in Pyatigorsk based at a mineral spring of the same name open in the same period. In 1821, on the order of P. Rikord, head of Kamchatka, the first baths for officers and soldiers of the Peter and Paul Garrison were built on the basis of the Paratunka springs.

In 1822, by order of Emperor Alexander I, a special commission was created to outfit the Caucasian mineral water resorts. On the recommendations of this commission, the Department of Home Affairs organized work on a detailed study of mineral water and on-site chemical studies of the water. For this purpose, A.P. Nelyubin (Fig. 4.1) was sent to the Caucasus; at that time, he was one of the most famous and well-established experts in the field of practical medicine and pharmacy, deputy director of the medical department of the Ministry of Home Affairs. Immediately on his arrival in the Caucasus, he organized a laboratory for chemical and physical research, which allowed him to conduct the most up-to-date research of that time.

Fig. 4.1 Alexander Petrovich Nelyubin. Lithograph from the portrait drawn by V.I. Pogonkin

In 1825, the two-volume work of St. Petersburg Medical and Surgical Academy professor A.P. Nelyubin *A Complete Historical, Medical-Topographic, Physico-Chemical and Medical Description of Caucasian Mineral Water* was published. The main part of this fundamental work is devoted to the chemical study of mineral water, both that which was already known and that discovered by the author himself.

During his study of springs, A.P. Nelyubin recorded their location, flow rate, water temperature, specific gravity, smell, taste, color, transparency, fumes emanating from water, and the effect of its exposure to air, heating, and cooling. He made the main emphasis of his research the accuracy of determining the content of gases (carbon dioxide and hydrogen sulfide) in mineral water. Nelyubin, with good reason, argued that the water of sources No. 17 and No. 18 in some respects outstrip water known in Western Europe, such as Vichy and Selters. In the valley of the Kislushi River, he found, studied, described, and numbered 23 sources, the names of which have been preserved to this day (Essentuki No. 1, No. 2, No. 3, No. 4, ... 17, etc.). These springs laid the foundation for the development of Essentuki as a national resort. Subsequently, Nelyubin conducted a comprehensive study of the Polyustrovsk mineral water and springs of Staraya Russa.

Nelyubin proved himself not only as a researcher, but also as a patriot. "This water has not only local significance, but also gigantic social significance for all of Russia," he wrote. "In Caucasus, every inch of the earth is watered with the blood of Russian soldiers, and the water we have received is important to us not only as a means of healing the wounds of Caucasian heroes, but also as a powerful means of introducing culture in local life" (Shevchenkoa and Shepin 2003).

In 1828, P.A. Savenkov, a professor at the Medical and Surgical Academy, published the book *Caucasian Mineral Water*, in which he gives detailed clinical recommendations for the treatment of various diseases using mineral water. These recommendations remain relevant today and are used as the basis for treatment in the resorts of the North Caucasus.

The active development of Caucasian mineral water during this period was even reflected in the work of poet A.S. Pushkin, who described the situation at that time (1831) in the passage in his poem *Eugene Onegin*:

...and, showing green, Mashúk,
Mashúk, of healing streams dispenser;
around its magic brooks
a pallid swarm of patients presses,
the victims, some of martial honor,
some of the Piles, and some of Cypris.
In waves miraculous the sufferer
plans to make firm the thread of life.
To leave the wicked years' offenses at the bottom
[plans] the coquette, and the old man
[plans] to grow young—if only for a moment.

In 1835, a course on toxicology and elucidation of mineral water was introduced at the Department of Medical Substances (Pharmacology) of Moscow University (General Charter of Russian Universities... 1835).

In 1837, on the order of General N.N. Muravyov-Karsky, the Saki branch of the Tauride (Simferopol) military hospital was opened. It played a special role in the course of the Crimean compaign, treating the wounded with water and mud from Lake Saki (Fig. 4.2). Since then, this place has been one of the leading balneological centers of the Crimean Peninsula and Russia, and the entire world, because the combination of the unique action of Saki Lake brine, local mud, and its own mineral water for internal use makes it effective even in cases of severely disabling diseases and injuries to the musculoskeletal and nervous systems.

The second half of the nineteenth century can be characterized as a period of significant development and establishment of the scientific foundations of balneology and drinking treatment.

In 1852, at the Department of Pharmacology of the Medical and Surgical Academy in St. Petersburg, on the initiative of G.K. Kullakovsky, the first independent course of balneotherapy (the study of mineral water) in Europe was organized.

In 1863, in Pyatigorsk, under the leadership of director of the Caucasian mineral water resorts S.A. Smirnov, the Russian Balneological Society was organized, the goal of which was to raise Russian resorts to the level of the best international ones. For this, the society published the first "Guide to Resorts of the Caucasian Group" and launched the weekly "Leaflet for Visitors to the Caucasian Mineral Water" (1863–1890). At this time, doctors working in the resorts of the Caucasus went on annual business trips abroad. All of this greatly enriched resort practice and made it possible to organize independent research.

In 1870, while digging a well, soldiers guarding the military road to Vladivostok discovered Shmakovka mineral springs, which later became the foundation of the

Fig. 4.2 Saki. Mud yielding. Photo from the archive of the Saki local history museum

largest balneological resort in the Far East. This is one of the only places in the world where the natural content of carbon dioxide in mineral water exceeds 3.5 g/l. Therefore, its effect, especially in external use, is many times greater than the effect of concentrated dry air carbon dioxide baths, and dosed respiration of air with a high content of CO_2 that evaporated from the mineral water is a powerful natural factor for stimulating the respiratory and circulatory organs.

Smirnov, being a staunch supporter of state resort management, in his book *Mineral Water in General and Caucasian Water in Particular*, published in 1884, wrote, "As part of the natural wealth of the country and having vast national importance, mineral water should naturally belong to the state" (Shevchenko and Shepin 2003).

Historical confirmation of these words may be seen in the intensive development of balneology after the nationalization of the resorts in 1917. In 1921, the Balneological Institute at the Caucasian Mineral Waters (in Pyatigorsk) was established. In 1922, the Tomsk Balneophysiotherapeutic Institute was established, and in 1921 the Central Institute of Balneology and Physiotherapy (now the FSBI National Medical Scientific Center for Restorative and Resort Medicine of the Ministry of Health of the Russian Federation) was opened in Moscow.

Currently, the organization includes a medical and scientific center in Moscow, which provides high-tech medical assistance for medical rehabilitation, as well as scientific and clinical branches, which provide medical assistance in the areas of "medical rehabilitation" and "resort treatment," located in Moscow Oblast, Krasnodar Krai, and Stavropol Krai.

4.2 History of Mud Studies

People began to use mud treatment (peloid therapy) a long time ago. There is evidence that people have used various types of mud to treat wounds and soothe skin irritations since prehistoric times. The first records of therapeutic mud use, written on clay tablets, were made in Mesopotamia in 2500 BC. In another riparian civilization, ancient Egypt, therapeutic mud was also actively used. Pharaoh's physicians used it as an anti-inflammatory and antiseptic. It is believed that even Cleopatra used healing clay masks to maintain her complexion. The Ebers Papyrus, dating from 1550 BC, which is considered one of the main surviving medical tracts from past centuries, describes the widespread use of healing clays for a wide range of diseases. In the ancient Greek period, peloids were used as an antiseptic for the treatment of skin diseases, and as a medicine for snakebites. Greek doctors also used treatment mud to fight malaria and gastrointestinal diseases (Carretero et al. 2006).

Mud therapy was especially popular in ancient Rome. Roman healers actively promoted the treatment with "fango" diluted volcanic mud for joint diseases, heart ailments, and mental disorders. In addition, in ancient times in Italy, they noticed the rapid healing of wounds on the legs of horses passing along mineral streams through mud springs.

In Europe, which is known for its balneological resorts, peloid therapy was widely used in the form of medicinal baths and wraps for rheumatic disorders, arthritis, gynecological and skin diseases, injuries, etc. Over time, peloid therapy even became fashionable. In the eighteenth and nineteenth centuries, people of the nobility sent their children to mud resorts in Sweden, France, Germany, and Austria, where a large audience gathered, not only to improve their health, but also to establish contacts in their professional fields and make a brilliant career. Many of these balneological resorts still use therapeutic mud, such as the Furnas Center located on San Miguel Island (Azores); Montecatini Terme, Ischia; and Abano Terme in Italy; Karlovy Vary in the Czech Republic; and Archena and El Raposo in Spain (Carretero et al. 2006). These are the most famous spa resorts, and they have existed for hundreds of years.

Therapeutic mud has a long history of use in Russia. One of the first mud springs, described by the ancient Greek scientist Pliny the Elder, is located in Crimea, in the north of the Kerch Peninsula. The unique Chokrak Salt Lake (it means "spring" in Turkic language) is often also called "silver," because its healing mud silt contains a high amount of silver ions (Fig. 4.3). In historical times, hydrogen sulfide springs and the salty water of the lake healed Scythians and Taurus, and Alexander the Great's soldiers improved their health here. In the Middle Ages, the healing mud of the lake was thought to give the Crimean khan miraculous powers of both combat and sexual potency. Since the nineteenth century, brine and mud from Lake Chokrak, saturated with iodine, bromine, magnesium, and other elements, have been exported to spa resorts in Italy and France.

Another famous place with mud treatment was the so-called Northern Resort located in the small town of Staraya Russa in Novgorod Oblast. In 1812, Dr. F.P. Gaaz, the discoverer of Caucasian healing water springs, came here to study the

Fig. 4.3 Chokrak Lake (Crimea)

mineral water springs, and in 1828, one of the best doctors in St. Petersburg, a physician in ordinary, doctor of medicine and surgery Georg Adolf Dietrich von Rauch came here as well. It was Rauch who remarked that the locals bathe all the time in the mud springs of the local salt lake, "in order to cure their diseases: ulcers, rubella, scrofula, arthritis" (cited by Dementiev and Dementieva 1990). As a result, he obtained the consent of Nicholas I to construct a resort. Special hospitals for soldiers and a mineral water institution were opened in Staraya Russa.

Thanks to the resort, Staraya Russa became one of the most comfortable cities in Europe. A railway station, paved streets, and a paved road connecting the city with St. Petersburg appeared. The first publications about the institution and its healing water were released. As Dr. M. Bukh wrote, "After wounds and other injuries, the effect of Staraya Russa's water is amazing," and Staraya Russa's therapeutic mud "is not inferior to any other, not only in Russia but also in Europe in general" (cited by Dementiev and Dementieva 1990).

Later, the mud bath resort was transferred to civilian use, and high-placed individuals began to come for treatment. As a result, Putevoy Palace was built in the town, where members of the royal family stayed. The recipe for a body wrap for society ladies has been preserved: the ladies' waists were pulled into a tight corset of therapeutic Staraya Russa mud, because it was believed that the mud actively "eats the hated fat folds on their sides" (Fig. 4.4).

In another part of the country, at an altitude of 700 m a.s.l., among the flowering steppe, on the border of the Kabardino-Balkar Republic and Stavropol Krai, lies the legendary Tambukan closed lake (Fig. 4.5), which is very small: 2 km long and 1

Fig. 4.4 Srednee Lake with treatment mud in Staraya Russa

Fig. 4.5 Tambukan Salt Lake (on the border of the Kabardino-Balkar Republic and Stavropol Krai) is a unique source of health. Along its coast, there are skeletons of dead trees, under its black surface, there is healing mud compressed by time and dead organisms

km wide. At first glance, it makes a "creepy" impression. Even in calm weather, the surface of the lake's water seems almost black, because the thick layer of extremely healing mud that covers the bottom is visible through it. Legends say that the famed Tamerlan took treatment baths and warmed his sore foot with the magical mud of this spring during his devastating invasion of the North Caucasus in 1395. However, recent archeological studies found that the healing properties of these mud were known even in the Bronze Age: excavations on the lake shore revealed traces not only of ancient settlements, but also bath houses near the lake.

Today, the name Tambukan is firmly attached to the lake, but earlier it was known as Tambi, which means "bad smell" in Turkic. That is one of the features of the sulfide silt treatment mud and the large mass of algae overlapping it. According to various estimates, there are about 1.5 million tons of healing mud at the bottom of the lake.

In the eyes of many researchers, the peloids of Tambukan Lake are several times more beneficial than the bottom sediments of the famous Dead Sea. The bottom sediments of the lake are a plastic silt mass belonging to the medium-sulfide highly mineralized group and containing many organic compounds and chemical elements: bromine, magnesium, silver, strontium, etc. It is with the treatment mud of this unique lake that the history of peloid therapy in the Caucasian Mineral Waters resort begins. The first chemical analyses of the lake's brine and silt were made in 1892. Treatment mud from the lake was already used in the hospitals of the Caucasian Mineral Water region in 1886 for gout, neurological diseases, and joint diseases. Then they started delivering it to Pyatigorsk and Zheleznovodsk. Treatment baths began to be used for patients: peloids were dissolved in mineral water, and the resulting "drenches" were used as a healing agent. In light of this, in 1915, a huge mud storage facility for 4,650 tons was built in Essentuki, and the largest institution for mud cures in Europe was opened. Today, it is one of the largest medical treatment and prophylactic complexes in Russia. During the Great Patriotic War, dressings with treatment mud saved many wounded soldiers who were treated in the hospitals of the Caucasian Mineral Waters, because the mud contains antiseptics that are similar to penicillin in their therapeutic effect.

Starting in the nineteenth century, institutions for mud cures were opened on the Baltic coast in Raushen (now Svetlogorsk), in Tinaki (1820) and Odessa (1820). Mud treatment began almost simultaneously in Slavyansk (1832), Sergiev Mineral Waters (1833), Staraya Russa (1839), and later in Lipetsk (1876), Caucasian Mineral Waters (1886), and Anapa (1900).

But Crimea remained one of the main places for peloid therapy. It was there in 1828 that the world's first mud resort was opened in Saki, and subsequently, 10 years later, a branch of the Simferopol Military Hospital was built. Mud treatment for patients wounded in the spine and spinal cord began in Saki during the Crimean War (1853–1856) at the suggestion of N.I. Pirogov. At the beginning of the 20th century, in 1910, a famous professor of resort spa treatment, S.S. Nalbandov, and the founder of neurosurgery in Russia, N.N. Burdenko, developed the scientific basis of peloid therapy and organized a hospital at the resort that was the first in Russia, where treatment was provided without charge. In the Saki resort named after Burdenko,

Fig. 4.6 The bath used in the Saki institution for mud cures in the nineteenth century

unique methods of mud treatment for patients with traumatic disease of the spinal cord were developed (Fig. 4.6).

Today, therapeutic mud is used in Russia in many resorts, and off-resort facilities number in the hundreds. Currently, in the Russian Federation, therapeutic mud of various geneses and all known types has been identified and used with good clinical effect: sulfide silt mud, sapropel, peat, and mud volcanic breccia. But since the early 1990s, the volume of geological exploration of natural treatment resources and scientific research have been gradually reduced due to the introduction of fundamentally new legislation on mineral resources since 1992, as well as changes in the general economic situation in Russia. Accordingly, the growth of proven reserves of mineral water and treatment mud has been negligible (Application of Peloid Therapy... 2015).

4.3 Cartographic Studies of Healing Springs

Healing water and mud resources have long been used in Russia as empiric remedies; however, their organized and scientifically based use began much later than in the European countries. In 1627, for the first time ever, information about hot springs appeared in the first gazetteer of the Russian lands, Book of the Great Sketch Map, which contains an inventory of springs on the "Pyatigorsk Circassian land" (Barandeev 1973).

The basis for the creation of balneological resorts was the Decree of Peter I of July 24, 1717 "On the search for mineral waters in Russia" for treatment of various diseases.

In 1817, the first cadastre of mineral springs in Russia was created under the guidance of academician V.M. Severgin and professor D.I. Sokolov. In Russia, several thousand mineral springs have been found. Relative accounting was carried out until the end of the 1980s at the mineral water intakes of trade-union health resorts, up to 75% of which are located in the European part of Russia, including the North Caucasus (Chernyaev 2000). By the 1990s, enterprises of the Ministry of Geology and other departments had compiled reviews of groundwater operational reserves for all economic regions of Russia, including mineral groundwater reserves (Operational reserves... 1992).

Currently, the State Balance of Mineral Reserves of the Russian Federation includes more than 1,000 deposits (deposit sites) of mineral groundwater (Mineral groundwater... 2018). They have been studied in most constituent entities of the Russian Federation. The number of deposits in the constituents varies from a single one (Chukotka Autonomous Okrug) to several hundred (Krasnodar Krai, Irkutsk Oblast). There are mineral water springs in Russia that are known throughout the world for their healing properties. The most significant mineral water types are concentrated in the Caucasus, North-Western and Siberian Federal Districts of the Russian Federation, and Primorsky Krai. There are world-famous water types: in Kislovodsk, Essentuki, and Zheleznovodsk—carbonated mineral water; in Sochi, Matsesta, Ust-Kachinsk, and Talga—sulfurous water; in Pyatigorsk, Belokurikha, and Uvildy—radon water; in Marcial, Polyustrovo, and other places, there is ferruginous mineral water.

In 2000, more than 400 mineral water deposits of various types were being exploited in Russia (Kulikov et al. 1991); however, there was practically no general account of deposits used and studied, nor any monitoring of their reserves. There is incomplete data about deposits with unconfirmed reserves operated by bottling plants, as well as departmental dispensaries.

Mineral water deposits are reflected in a number of overview maps – Map of thermal water of the USSR (Mavritsky 1965); Map of medicinal mineral water in the USSR (Ivanov 1968); Map of mineral water of the USSR (Yarotsky 1984); Map of forecast resources and operational reserves of the USSR's main mineral water types (Vartanyan 1984), etc. and atlases—Atlas of hydrogeological and engineering-geological maps of the USSR (Rogovsky 1983); Atlas of maps of thermal water resources in the USSR (Mavritsky 1985) (Figs. 1.1 and 1.2). Russian National Research Institute of Hydrogeology and Engineering Geology is creating GIS Atlas of hydrogeological maps, including maps of mineral and thermal water in the Russian Federation (Lukyanchikov et al. 2014).

The recent National Atlas of Russia (2007) includes two overview maps—"Provinces of Mineral Water" (scale 1:30.000.000) and "Types of Mineral Water" (scale 1:75.000.000).

Maps showing the distribution of mineral water sources are contained in a number of regional atlases, such as the most informative maps in Atlas of Zabaykalie (1967) (Fig. 1.3a, b), and Atlases of Republics of Adygea (2001) and Tatarstan (2005) (Fig. 1.4), and others. The most informative maps are in the Atlas of Zabaykalie.

Informational directories from the Federal Water Resources Agency for Basin Water Management Boards contain textual information about the resources and the quality of groundwater. The multivolume Russian Natural Resources Ministry publication Water of Russia: Water-Resource Potential and State reports contain information on mineral, thermal, and commercial groundwater. In the Guide albums Water Resources published in the constituent entities of the Russian Federation, there is information about healing springs (Chernyaev 2000; Penza Oblast... 2013; Ivanovo Oblast... 2015; State report... 2017, and others).

Mud deposits in the Russian Federation have been thoroughly studied and displayed on maps—Map of the therapeutic mud of the USSR (Ivanov 1968); Catalog of the mud deposits of the USSR, (1970); Map of therapeutic mud deposits of the USSR (Ivanov et al., 1968) (Fig. 1.5). Of the 500 known mud deposits, only a quarter of them are used for medicinal purposes in health resorts and in non-resort settings.

Quite detailed information about springs and mud deposits is provided on tourist maps of individual regions and health resort establishments, as well as in descriptions of tourist site attractions that can be found online in bountiful numbers.

References

Application of Peloid Therapy in treatment and rehabilitation programs. Clinical recommendations. Approved at the XIII International Congress “Rehabilitation and spa treatment”. Moscow. 2015 (in Russian)

Atlas of the Republic of Adygea (2001) Maykop: Laboratory of geoinformation technologies of Adyghe State University, 77p (in Russian)

Atlas of the Republic of Tatarstan (2005) Moscow: PKO Kartografiya, 120p (in Russian)

Atlas of Zabaykalie (1967) Moscow-Irkutsk: GUGK, 176p (in Russian)

Barandeev AV (1973) Book of the Great Sketch Map as a source of toponymical research. Izvestiya Akademii Nauk SSSR. Seriya Geograficheskaya i Geofizicheskaya 5:136–143 (in Russian)

Carretero MI, Gomes CSF, Tateo F (2006) Clays and human health. In: Bergaya F, Theng B, Lagaly G, editors. Developments in clay science. Vol.1. Handbook of clay science. Amsterdam: Elsevier Science, pp 717–741

Catalogue of the mud deposits of the USSR: an appendix to the map of therapeutic muds of the USSR, scale 1: 8.000.000 (1970) Moscow: CNII kurortologii i fisioterapii, 131p (in Russian)

Chernyaev AM (ed) (2000) Water of Russia. Water-resource potential. Yekaterinburg: AQUA-PRESS, 420p (in Russian)

Complete collection of laws of the Russian Empire. Collection 1713-1719. Vol. V, pp 498–499, 684–687 (in Russian)

Dementiev KI, Dementieva MA (1990) Written word about Staraya Russa. Novgorod: The State Novgorod Museum-reservation, 199p (in Russian)

General Charter of Russian Universities dated 26 Jul 1835. (1835) In: Complete collection of laws of the Russian Empire. Collection 1825-1881. Vol. X), 8337p (in Russian)

Ivanov VV (ed.) (1968) Map of mineral healing waters of the USSR (1:4000000). Moscow (in Russian)

Ivanovo oblast (2015) Water resources. Nizhny Novgorod: Avanta-Print, 100p (in Russian)

Kulikov GV, Zhevlakov AV, Bondarenko SS (1991) Mineral medicinal waters of the USSR: Book of reference. Moscow: Nedra Publishers, 399p (in Russian)

Lukyanchikov VM, Baron VA, Golitsyn MS, Plotnikova RI, Tikhonenkov YuE (2014) Modern hydrogeological map of the Russian Federation (1: 2500000). Sovremennaja-gidrogeologicheskaja-karta-rossijskoj-federatsii-141909.htm. Accessed 5 Mar 2019 (in Russian)

Mavritsky BF (ed) (1965) Map of the thermal waters of the USSR (1:2500000). Moscow (in Russian)
Mavritsky BF (ed) (1985) Atlas of thermal water resources of the USSR (1:10000000). Moscow (in Russian)
Mineral groundwater (2018) In: State report "On the state and use of mineral resources of the Russian Federation in 2016 and 2017". Moscow: Ministry of Natural Resources and Environment of the Russian Federation, pp 330-331 (in Russian)
National atlas of Russia (2007) Vol. 2. Nature and ecology. Moscow: GOSGISCENTR, 495p (in Russian)
Operational reserves of groundwater of the RSFSR approved by the State Reserves Committee of the USSR (TKZ) as of 01.01.1991 (1992) Moscow: Rosgeolfond, 1044p (in Russian)
Penza oblast: water resources (2013) Nizhny Novgorod: Delovaya Poligrafia (in Russian), 100p (in Russian)
Rogovsky NV (ed) (1983) Atlas of hydrogeological and geotechnical maps of the USSR (1:5000000–1:15000000). VSEGINGEO (in Russian), Moscow
Sergeenko ME (1960) Life of Ancient Rome. Moscow-Leningrad: Nauka, 368p (in Russian)
Shevchenko YuL, Shepin OP (eds) (2003) Resorts of the Caucasian Mineral Waters as a national treasure: formation and development (1803-2003). Moscow, 184p (in Russian)
Starikov SM (2018) On medical geography of healing springs used in sanatorium organizations of the Ministry of Defense of Russia. Voyenno-meditsinskiy zhurnal 339(1):94–96 (in Russian)
State report "On the condition and use of water resources of the Russian Federation in 2016" (2017) Moscow: NIA-Priroda, 300p (in Russian)
Vartanyan GS (ed) (1984) Map of forecast resources and operational reserves of the main mineral waters of the USSR (1:5000000). Moscow (in Russian)
Yarotsky LA (ed) (1984) Map of mineral waters of the USSR (1:2500000). Moscow (in Russian)

Chapter 5
Health Resort Complex of Russia

Abstract The chapter provides a general description of the health resort complex in Russia. Health-improving areas and resorts in the country have almost all the types of healing mineral water, therapeutic mud, and other natural healing resources known in the world. Selected descriptions of the resorts, their history, and the current state are presented. The text is accompanied by maps of resorts and health areas distribution. The focus of this chapter is on the oldest resorts in Russia–Marcial water, Staraya Russa, etc., as well as on the most famous resorts of the Caucasian Mineral Waters and less well-known resorts of Transbaikal and other regions.

Keywords Health resort · Health areas · Healing resources · Russia

5.1 General Characteristics of the Health Resort Complex

The Federal Law "On Natural Healing Resources, Health-Improving Areas and Resorts" (1995) defines principles of state policy and regulates relations in the study, use, development, and protection of natural healing resources and health-improving areas.

In accordance with federal law, the resort fund is the totality of all of those declared or considered natural healing resources, health-improving areas, as well as resorts and resort regions (districts).

Health-improving areas and resorts of the Russian Federation have almost all the types of healing mineral water, therapeutic mud, and other natural healing resources known in the world and can ensure the preservation and further development of the health resort complex of the Russian Federation (Fig. 5.1).

Natural-healing resources, health-improving areas, and resorts are the basis of the health resort complex and are specially protected objects and territories that have special aspects of their use.

State registration of the resort fund of the Russian Federation and the State register of health-improving places and resorts, including health resort organizations, is carried out by the Ministry of Health of the Russian Federation as part of the powers established by the legislation of the Russian Federation.

S. Malkhazova et al., *Healing Springs of Russia*,
https://doi.org/10.1007/978-3-030-83534-7_5

Fig. 5.1 Solvychegorsk Resort in Arkhangelsk Oblast is located in a picturesque site near Vvedensky Monastery (now inactive). For medicinal purposes, bromine, chloride-sulfate-sodium, and hydrogen sulfide mineral water is used, as well as silt mud

According to the State Register of the Resort Fund of the Russian Federation, there are currently more than 100 documents on the recognition of territories as health-improving places or resorts, including demarcation of districts of sanitary (mountain) protection in the database of normative legal acts.

According to the State Register of the Resort Fund of the Russian Federation, 44% of health resort organizations are state property, 3% belong to municipalities, and 53% are private property.

In 2016 and 2017, among 495 state health resort organizations, 140 were federal organizations and 355 were regional organizations.

The largest number of state health resort organizations is concentrated in the Republic of Crimea, Krasnodar Krai, and Stavropol Krai.

The goal of the Development Strategy for the Health Resort Complex of the Russian Federation (2018) is to increase the availability of health resort treatment by creating a modern health resort complex that ensures the development of health-improving areas and resorts, effective use of natural healing resources, development of the material and technical base of health resort organizations, and realization of the health tourism potential of Russia.

Below are listed resorts (health-improving areas) and health resort organizations according to the list from the program-informational complex of the State Register in 2018 (Map 5.1 and 5.2).

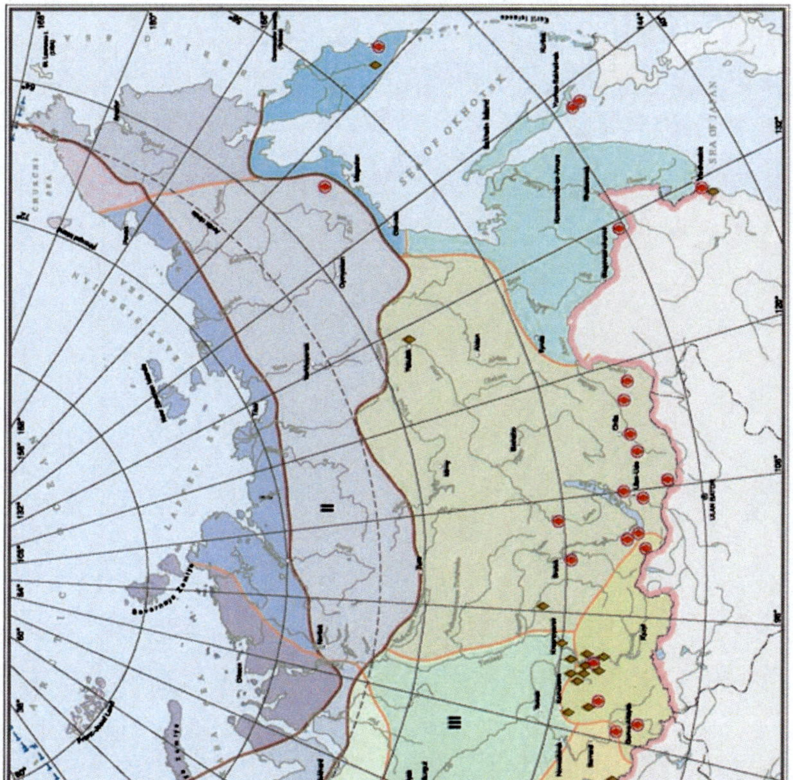

Map 5.1 Resorts and health areas

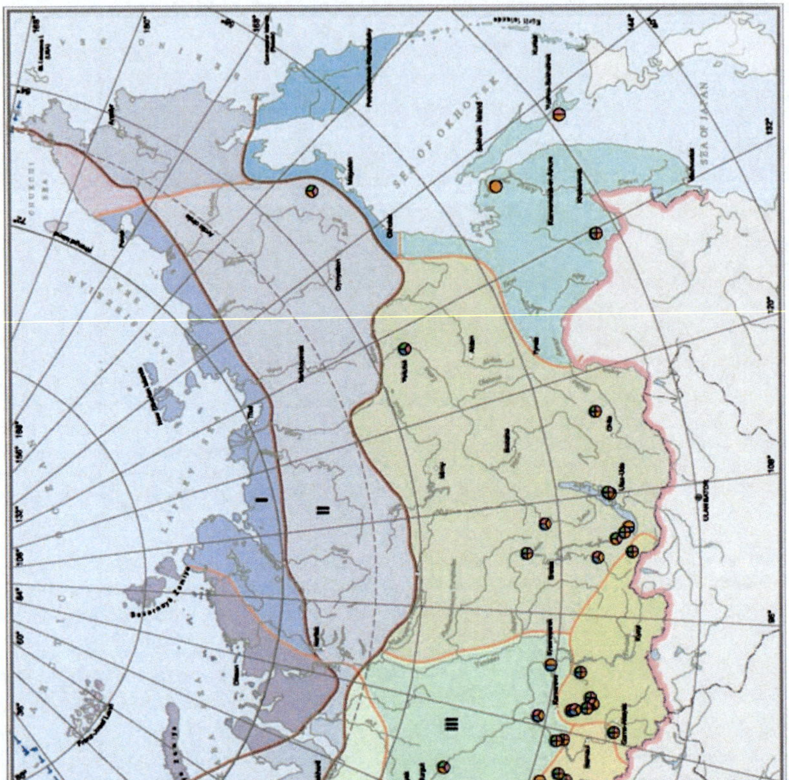

Map 5.2 Health resorts using healing water and mud

5.2 The Longest-Standing Russian Resorts

5.2.1 The First Russian Resort–Marcial Water

The first Russian resort, Marcial water, was established by Peter the Great in the valley with ferrous mineral springs, located 50 km from Petrazodovsk (capital of the Republic of Karelia). From 1719 to 1724, Peter the Great was here four times and spent a total of about 90 days at the resort. Especially for the tsar and his guests there were built wooden palaces. Unfortunately, they have not survived to the present day. But there is the church of Apostle Peter, the one and only architectural monument that remained in Karelia of the Petrine era.

Currently, the resort has two large health-improving complexes: the Marcial Water spa resort functioning since 1964, and the Dvortsy spa resort opened in 2003. Their main healing factors are ferruginous sulfate magnesium-calcium-sodium water, and therapeutic mud—high-sulfide and medium-sulfide freshwater sapropels from the bottom of nearby Gabozero lake. In addition to resort vacationers, more than 20 thousand tourists come here from all over Russia and the world every year to visit monuments of the Petrine era (Kapusta 2006).

Mineral water spills from four wells, with a depth of 7–14 m. Water has different mineralization, and contains a different amount of dissolved iron. According to a study conducted by the Medical Rehabilitation and Balneology RSC in 2016, its content is 37–64 mg/l. In addition to iron, the water in these springs contains a various amount of sulfates and chlorides, magnesium, sodium, calcium, and metasilicic acid. Different mineral content and composition allow differentiated use of water from these four springs. Mostly, this differentiation is performed as follows:

Spring No. 1 water is used for diseases of the genitourinary system (category XIV), including gynecological pathology, urolithiasis, as well as metabolic disorders (category IV).

Spring No. 2 water is used for diseases of the digestive system (category XI): chronic gastritis, chronic cholecystitis, and for endocrine, nutritional, and metabolic diseases (category IV).

Spring No. 3 water is used for diseases of the circulatory system (category IX), diseases of the musculoskeletal system and connective tissue (category XIII), and diseases of the nervous system (category VI).

Spring No. 4 water is used for hypoferric anemia, hemopathy, hematopoietic diseases, and diseases of the blood and blood-forming organs and certain disorders involving the immune mechanism (category III), as well as for endocrine, nutritional and metabolic diseases (category IV).

Marcial mineral water's complex effect on homeostasis and metabolism is due to two main mechanisms. Firstly, it is hemoglobin synthesis activation and increasing oxygen transportation, which increases energy transfer, microcirculation, and other processes in almost all organs and tissues and, as a result, leads to an increasing functional state of the main body systems (cardiovascular, respiratory, nervous, digestive, musculoskeletal, endocrine, genitourinary, etc.). Secondly, a nonspecific reaction to the low-mineralized water with unique physicochemical properties activates adaptation mechanisms, stimulates immune responses, and it leads cumulatively to a general tonic effect. Thus, marcial mineral water has universal healing properties that can treat many diseases, functional disorders, and pathological conditions of the human body. And their 300-year successful experience of use is the most reliable confirmation.

5.2.2 The Staraya Russa

The geographical location of Staraya Russa, one of the longest-standing Russian cities, which already existed in the second half of the twelfth century, is associated with a salty mineral water deposit (Fig. 5.2). For a long time, locals were engaged in salt production. In the fifteenth—nineteenth centuries, before the mass development of rock salt production, the city was one of the largest industrial and commercial

Fig. 5.2 Staraya Russa—Slightly alkaline (pH 7.4) mineral water output to the surface, Muravyov fountain (**a**); mineral water outcrop (**b**); Non-freezing Solonik on the Polest River (**c**)

Russian centers. Due to its strategic and economic importance, Staraya Russa was visited by Peter the Great (1693, 1724); Catherine II (1780); Alexander I (1823), who signed a decree on the introduction of military settlements in Staraya Russa in 1824; and Nicholas I (1826, 1827, 1830), who approved the "Mineral Water Establishments in Staraya Russa" project in 1830.

Officially, 1828 is considered the foundation year of the balneological resort in Staraya Russa, when, after an inspection of military hospitals located near the city at that time, a physician in ordinary G. Rauch studied the use of the healing water and mud of Staraya Russa and made a report to war minister A.I. Chernyshov, who decided to start the construction of the first medical buildings for soldiers and officers. Well-known professors F.P. Gaas (Moscow) and A.P. Nelyubin (St. Petersburg) took an active part in the study of the composition and action of the mineral water of Staraya Russa. A review article, "Observations on the Medical Properties of Staraya Russa Salt Springs" was published by M.P. Lamovsky, the senior doctor of the old Russian military hospital, in the Military Medical Journal in 1837.

In 1857, the resort was visited by the deputy chairman of the Imperial Russian Geographical Society and minister of state property Count M.N. Muravyov. On his order, for the needs of the resort, drilling was begun for a new well in 1858 and ended in 1859 with the opening of the spring named in honor of the count. The image of the existing Muravyov fountain, which has changed appearance repeatedly, but consistently delivers mineral water at the natural pressure from a depth of 118 m, has become the hallmark of the resort (Korovina 2015).

When leaving the well, water has almost unchanged characteristics: about 20 g/l of salt and a constant temperature of +11 °C. There are seven mineral springs in the resort, and the water from them is used for medicinal purposes and enters a cascade of artificial mineral lakes, some of which are used to receive and regenerate therapeutic sulfide-silt mud. About 10,000 m^3 of mineral water passes through the lakes daily (Korovina 2015). Then the water flows into the Polest River, forming a non-freezing stream called Solonik by the locals.

In the Staraya Russa resort, mineral water is used for therapeutic purposes in the form of baths, rinses, irrigation, and therapeutic pools, as well as for drinking treatment. The chemistry of drinking mineral water is: spring No. 11 is medium mineralized, chloride sodium-magnesium-calcium, slightly alkaline; spring No. 12 is low mineralized, chloride calcium-magnesium-sodium, slightly alkaline.

The main indications for the use of mineral water are musculoskeletal, neurological, respiratory (not of tuberculous etiology), gastrointestinal, metabolic, circulatory, urological, and gynecological diseases.

In the Sobornaya Storona area of Staraya Russa, there are two active, but unused, salt springs. These springs were used for the state salt factory that closed in 1869. Work is currently underway to design a treatment and rehabilitation center based on them.

The Staraya Russa resort has a rich history of use of many natural healing factors: climate (the average temperature is +18.4 °C in the summer, and –7.5 °C in the winter), mineral water (internal and external use), and therapeutic mud. A hydrogeological monitoring and observation laboratory and the country's first mud farm operate at the resort.

5.2.3 *Solonikha Resort in Arkhangelsk Oblast*

The Solonikha resort is located in the south of Arkhangelsk Oblast (Fig. 5.3), in the Krasnoborsky district, far from large cities and industrial centers, in the valley of the Evda River, a tributary of the Northern Dvina, in a forest of centuries-old pines. The emerald salt spring lake gives the district its extraordinary character.

The history of Solonikha is rooted in the old days and was first mentioned in 1626. Three villages (Lukinskaya, Ignatovskaya, and Penye) were known in the region for their salty spring, the healing water of which was used to treat locals, especially for aches in their arms, legs, and lower back. Czar Ivan the Terrible gave the land in the southern part of the Northern Dvina to Stroganov merchants, so that they would supply the state with salt. Four salt springs were discovered in this area. The most profitable was Solonikha spring: "a pound of salt was brewed from 125 litres of its water."

The Solonikha resort was organized thanks to tremendous efforts by the artist Alexander Alekseevich Borisov. According to legend, Borisov was seriously injured as a child, which deprived him of the ability to move around. The future artist was literally saved by the living water of Solonikha.

The health resort began with one bathing section. Its opening took place on July 27, 1922. The first doctor in the resort was Dinora Alexandrovna Fabrikant, and Borisov, who had invested a lot of effort and personal funds in the construction, was called the chief manager at the time.

Fig. 5.3 Territory of the Solonikha resort

In 1927, the first dormitory was built according to a design by Borisov. To this day, it remains a monument of architecture and the real pride of Solonikha.

Important healing factors at Solonikha resort are mineral water and peat therapeutic mud. The resort operates two springs of mineral water:

- sulfate–chloride-sodium water with mineralization of 22.2 g/l, which is used externally (in the form of baths, inhalations, irrigation); clear water, without sediment, with a bitter-salty taste, odorless;
- sulfate–chloride-sodium water with mineralization of 5.37 g/l, which is used for intake; the water is clear, without sediment, with a salty taste, odorless.

A well with drinking water was drilled in 1964, its depth is 255 m; water temperature is +40 °C.

The Solonikha resort has its own deposit of therapeutic peat mud.

5.2.4 Healing Springs of Khilovo: Water and Mud of Pskov Oblast

Khilovo was first mentioned in Novgorod scribal books in 1539. They said that a man with the old Russian name Khil came here and the name of the village came from him.

In 1805–1807, Khilovo was divided into two estates: Khilovo Balavinsky and Khilovo Korsakova. Peter Mitrofanovich Balavinsky settled in the first of them in 1819. Having retired with the rank of captain, he took up a household and raised his sons. The eldest, Peter, entered military service, and visited various places in Russia and abroad. Arriving in Khilovo in 1865, he ordered the Chernaya River to be diverted from its course, at the bottom of which hydrogen sulfide springs were discovered.

According to legend, Peter Balavinsky's cook found the healing spring and cured his sore feet. News of this miraculous healing spread throughout the district and reached St. Petersburg.

In the same year, the German master of pharmacy Kasselman arrived in Khilovo. He was the first scientist to take samples, make analyses, and give his opinion on the healing power of the Khilovo water.

In 1866, professor of chemistry and famous Russian composer A.P. Borodin came here. He deduced the detailed formula of Khilovo's hydrogen sulfide water and admired how good its ionic and mineral composition was for treating sickness. The scientist presented the results of the research to the military department.

Based on Borodin's conclusions, the military department signed a contract with Peter Balavinsky to rent his land and build a resort to treat sick soldiers and officers.

Gradually, civilians began to arrive in Khilovo too. A water and mud bath building was built in the form of a U with male and female divisions, steam baths, and a shower, as well as a division for treatment with milk and whey. Near this building, Balavinsky erected a stone on which the family crest and motto, "Benefit is above victory," is carved along with the resort's year of founding, 1865.

On May 9 (22), 1889, an imperial decree was issued in the name of the minister of state property with an order "to take measures to protect the Khilovo springs from spoilage and exhaustion". With that, the springs gained national significance.

At the beginning of the twentieth century, there was an active interest in the health resort (Fig. 5.4). There were plans for its reorganization, and even the question was raised of making Khilovo one of the top ten best European resorts.

In 1920, the resort received a letter of protection signed by the People's Commissar of Health of the RSFSR N.A. Semashko. According to it, the resort was a medical institution of national importance.

In 1932, specialists from the Institute of Resort Medicine arrived to study Khilovo. They made a discovery: a unique therapeutic sapropelic mud was discovered in the local Lake Lunevo.

During the Great Patriotic War, the Germans placed a hospital on the resort's territory, and as they were leaving, they destroyed the water and mud bath building and blocked the wells. As soon as Khilovo was liberated from the invaders, the locals set about restoring the resort. In 1947, the first patients arrived for treatment. Since that year, the resort has not ceased activity even for a single season.

In 1959, mineral water extracted from the resort's unique natural springs was discovered. Its quality was highly appraised and it earned a gold medal at the International Exhibition of Mineral Water in Milan, Italy, in 2001.

Fig. 5.4 Sanatorium Khilovo (Pskov Oblast)

By a resolution of the Council of Ministers of the RSFSR of January 6, 1971, the Khilovo resort was included in the list of resorts of republican significance.

5.2.5 Kashin, the Famous Resort of Tver Oblast

The Kashin deposit of mineral water is the most extensive and highly developed in Tver Oblast.

By the seventeenth century, the natural springs in Kashin had been known to the locals for their healing properties for several centuries. They were called "sacred" and "God-given." Pilgrims who came to Kashin to worship St. Anna Kashinskaya used water from three springs on the banks of the Maslyatka River to treat a variety of diseases.

In 1808, the district doctor, D. Chernyavsky, compiled a detailed "Description of the Mineral Springs Located in Kashin." He contributed to the improvement of the surrounding area and the construction of a house for visiting patients. Kashin water gained some fame, it was included in Villier's "Military Pharmacopoeia" and studied by Professor Reis. But gradually, interest in the springs decreased, and by 1825 they were in a state of neglect.

In 1884, Dr. A.V. Alekseevsky reopened the treatment season at the mineral water. New log houses were made, tents for bathing were set up, and a garden with gazebos and a pavilion was built. Water samples were provided to Professor Lachaninov, and he noted that they were richer than the mineral water of Zheleznovodsk. Therapeutic indications for water treatment have been developed. The subsequent chronicle of the resort is filled with the following events:

In 1892, Emperor Alexander III, by a High Decree, affirmed the social significance of the Kashin mineral springs.

In 1901, mining engineer Kolsky drilled the first mineral well in the spring area.

In 1915, the first hydropathical establishment was built (Fig. 5.5).

In 1949, the development of the modern hydromineral base of the Kashin resort and the drilling of deep mineral water wells was begun.

Fig. 5.5 Kashin—the first hydropathical establishment, 1915 (**a**); modern spring (**b**)

5.2.6 Lipetsk—An Ancient Russian Resort

Peter the Great is considered the legendary discoverer of the Lipetsk mineral springs (Zhirova 2018). At the end of the seventeenth century, during the Azov campaigns, Tsar Peter, visiting Lipetsk, noted two existing factors: the presence of ferruginous stone and ferruginous mineral water, and he experienced its healing power.

"Simultaneously with the foundation of a factory by Emperor Peter I in 1700, springs of healing mineral water were opened in Lipetsk, for which a palace was built for the royal visit. The Tsar personally used the open water when he visited Lipetsk," wrote Dr. N.N. Maksheev in his guide "Lipetsk Mineral Water and Mud Baths," published in 1904.

The story goes that Peter the Great was not only treated with Lipetsk water himself, but also instructed the people: "Whoever uses this water will absolutely receive healing from various cruel diseases" (Narcissov and Narcissov 2018).

By the end of the eighteenth century, Lipetsk water was very famous, patients from different places flocked there. Even then, according to contemporaries, "up to 300 visitors and surrounding residents gathered annually for water drinking" near the healing springs.

The nineteenth century can be called a golden page of the history of Lipetsk. It was the time of formation and development of one of the first resorts in Russia. In 1803, the resort season was opened. Lipetsk water attracted the attention of scientists. Soon treatment "on the water" gained wide popularity. Every season, people wishing to improve their health began to come here from St. Petersburg, Moscow, Tambov, and Voronezh. Then the government decided to allocate funds from the state treasury to outfit the resort. Coverage about Lipetsk water in periodicals made it famous not only in Russia, but also in Europe. Considering the great popularity of the resort, the government made an official decision to open a state resort here called Lipetsk Mineral Water. A decree on the establishment was signed in 1805 by Alexander I (Zhirova 2018).

Construction of the resort began (Fig. 5.6). There was an increased focus on the medical part of the resort: new springs were discovered, treatment with ferruginous mud was introduced, etc. According to some reports (Narcissov and Narcissov 2018), the resort was designed by the famous architect and park builder A.N. Voronikhin (designer of the Kazan Cathedral in St. Petersburg and one of the creators of the Peterhof and Pavlovsk architectural ensembles). The Lower and Upper "English" gardens were laid out, and a building for mineral baths was built according to his design. There were plans to build a spa hall; over Peter the Great's spring, a beautiful pavilion with forty columns was erected. That masterpiece of classical architecture stood until 1924.

The resort's medicinal means (mineral water, which was compared in its healing properties with the water of the well-known foreign springs of Permont, Spa, Schwalbach, and Rosenbrunen, and later the ferruginous mud, for which Lipetsk was called "Russian Franzensbad") were famous throughout Russia. The resort season began in mid-May and ended in late August. Lipetsk was visited by nobility from the capital at the time, and there were well-known writers and actors, and commoners from

Fig. 5.6 Bath building of the early twentieth century

neighboring provinces. In wartime, wounded soldiers and officers were treated here (Zhirova 2018).

The names of many famous people are associated with Lipetsk mineral water: statesmen, writers, musicians, etc. At different times, the resort was visited by such prominent people in Russia as Peter the Great, Alexander I, Alexander II, G. Derzhavin, V. Zhukovsky, N. Karamzin, A. Griboedov, A. Bunin, S. Zhikharev, and A. Shakhovsky, who wrote the play "Lipetsk Water," in which the city was described by one of the characters: "Lipetsk is an earthly paradise." There were numerous relatives of Pushkin, the Goncharovs, Anna Kern, relatives of M. Lermontov (including his father), Arsenyev, P. Bartenev, P. Boborykin, Y. Polonsky, G. Uspensky, N. Rubinshtein, and many, many others.

The bathing room building, constructed directly by architect A. Voronikhin, the oldest building in the city, has survived to our days in a reconstructed form (Narcissov and Narcissov 2018). One of the oldest Peter the Great monuments in the city, built in 1839, is also associated with mineral water. It is located in the middle of Petrovsky Spusk. It was built with funds from St. Petersburg merchant Pavel Nebuchenov, who was cured of his ailment by Lipetsk mineral water.

In Soviet times, full reconstruction of the Lipetsk resort began in 1962. At present, the resort's hydromineral base is artesian mineral water wells. These wells produce chloride-sulfate sodium water. Of the microcomponents, bromine, fluorine, boron, and copper were found. Lipetsk mineral water is similar in its chemistry to Pasha Tele mineral water in Feodosia.

The mineral water has a therapeutic effect on the human body because of the whole complex of its physical and chemical properties (Fig. 5.7). Lipetsk mineral water is successfully used to treat a number of gastrointestinal diseases, which has been confirmed in a number of clinical and experimental observations. It tastes good,

Fig. 5.7 Lipetsk mineral spring

quenches thirst, and improves metabolic processes in the body. With systematic use (25–30 days), it normalizes functions of the stomach, liver, gall bladder, intestines, and pancreas.

Lipetsk mineral water is characterized as water with low mineralization (Fig. 5.8). The main indication for the internal use of Lipetsk water is gastritis, colitis, hepatitis, cholecystitis, peptic ulcer, and metabolic disorders. The water changes the activity of the gastrointestinal system, and gently stimulates its secretory and motor functions.

Fig. 5.8 Spring of drinking water

5.2.7 *Resorts of the Caucasian Mineral Waters—A Healing Region of Russia*

Resorts of the Caucasian Mineral Waters (CMW) are represented by the famous Kislovodsk, Essentuki, Pyatigorsk, and Zheleznovodsk resorts of federal significance, as well as the Kumagorsky resort with local significance (Krivobokov 1994; Sheiko and Manshin 2011). Treatment at CMW resorts is based on the use of various types of natural mineral water, healing mud, as well as the landscape and climatic conditions of the region (Map 4.1) (Schaefer 2017).

Kislovodsk resort. It is located in the south of the CMW district, in a distinct basin in the valley of the Berezovka, Olkhovka, and Podkumok Rivers, at an altitude of 800–1,160 m. The health resort base consists of 38 resorts with a capacity of 8,861 beds, Oktyabrsky and Yelava Narzan baths, and a small institution for mud cures. Mineral water for drinking treatment is given out in three resort galleries (Narzan spring gallery, Round well room, and Zhelyabovsky spring). Many resorts have their own bathing sections with a supply of mineral water via stainless steel mineral water pipelines (Fig. 5.9). Some resorts have their own drinking well rooms. The main area of spa resorts treatment is cardiovascular and respiratory diseases. In the late 1980s, more than 200,000–250,000 people took treatment and rested at this resort annually. Currently, the number of resort visitors reaches 100,000–150,000 per year.

Fig. 5.9 Kislovodsk. Hydropathical establishment

Essentuki resort. It is located in the central part of the CMW district, in the Podkumok River valley, at an altitude of 600–650 m. The main medical area of the Essentuki resort is gastrointestinal diseases and metabolic disorders. The health resort base is represented by 16 resorts with 4,977 beds, two hydropathical establishments, an institution for mud cures, and four general spa drinking galleries (well rooms), where Essentuki №17, Essentuki №4, and New Essentuki mineral water types are dispensed.

The hydro-mineral base of the Essentuki resort is 751 m^3/day of various types of mineral water. It can provide for 6,000–7,000 people simultaneously, or 70,000–90,000 a year. In the 1980s, more than 100,000–150,000 people a year were treated at the resort. Currently, resort visitors do not exceed 70,000–90,000 per year. No more than 30–40% of the reserves of the available hydromineral resource are used.

Pyatigorsk resort. It is located in the eastern part of the CMW area, on the left bank of the Podkumok River, at the foot and on the slopes of Mount Mashuk (Fig. 5.10), at an altitude of 510–630 m. Due to the presence of various types of mineral water and therapeutic mud of Tambukan Lake, the Pyatigorsk resort is multi-specialized. It treats musculoskeletal, skin, gastrointestinal, urological, gynecological, and other diseases. The health resort base has 17 resorts for 3,560 beds, 5 general hydropathical establishments, an institution for mud cures, and 8 drinking well rooms with carbonic and salt-alkaline (Essentuki-type) mineral water. At the end of the twentieth century,

Fig. 5.10 Pyatigorsk. Popular baths

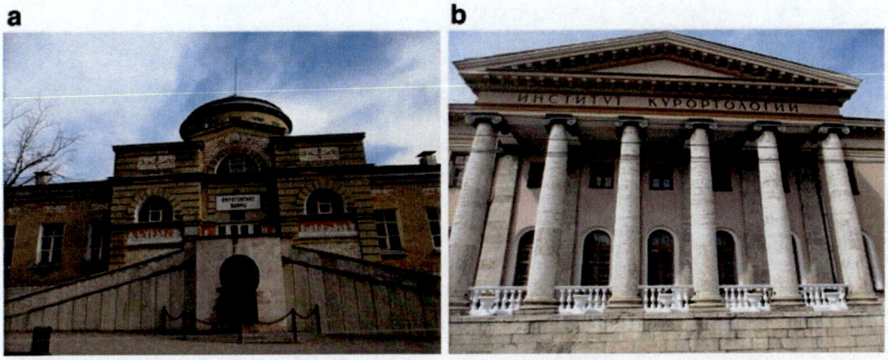

Fig. 5.11 Pyatigorsk-Pirogov baths (**a**), Institute of Resort Medicine (**b**)

more than 220,000–250,000 people a year were treated at the resort. Only 7–10% of the available mineral water reserve is currently used. Vacationers do not exceed 90,000–100,000 people per year. On average, not more than 0.7–0.9 m^3 of mineral water is applied to the treatment process at the resort, with a norm of 3 m^3 per vacationer.

At the Pyatigorsk resort, the significant resource of mineral water and therapeutic mud can provide highly effective treatment more than 200,000–250,000 people annually. According to the Pyatigorsk State Institute of Resort Medicine, for more complete development of hydro-mineral resources, it is necessary to increase the capacity of resort facilities by two to three times (Fig. 5.11).

Zheleznovodsky resort. It is located in the northeastern part of the CMW area, on the southern slopes of Zheleznaya Mountain, at an altitude of 600–650 m. (Fig. 5.12). The resort specializes in the treatment of genitourinary and gastrointestinal diseases and metabolic disorders. The health resort base is represented by 14 resorts with 3,447 beds, four resort-wide drinking well rooms, a hydropathical establishment, and an institution for mud cures. At the end of the twentieth century, more than 100,000 people a year were treated at the resort. Currently, resort visitors do not exceed 50,000–80,000 per year. No more than 30–40% of the reserves of the available hydromineral resource of the resort are used.

Kumagorsk Regional Rehabilitation Hospital (Kumagorsk resort). It is based on a mineral water deposit of the same name. The Kumagorsk deposit of hydrogen sulfide (sulfide) water is located 12 km northwest of the Mineral Waters and is confined to the dome-shaped uplift of Mount Kokurtly, the northernmost laccolith of the CMW region. By its chemistry, Kumagorsk mineral water is close to the sulfide water of the Goryachy Klyuch resort in Krasnodar Krai. It is successfully used for balneotherapy. Its current use does not exceed 20% of reserves. About 3,000 people are treated at the Kumagorsk hospital annually.

The dynamics of visiting CMW resorts at the beginning of the twenty-first century shows that the number of vacationers has stabilized at the level of 480,000–500,000

Fig. 5.12 Zheleznovodsk. Smirnovsky spring

a year. There is a steady decline in the consumption of mineral water for spa resort treatment, as their average amount decreased from 1.6 m^3 per person in 2000 to 0.7–0.9 m^3 per person in 2012. The standard requirement for a 21-day stay at the resort is 3.0 m^3 per person based on 8–10 baths with mineral water, irrigation, flushing, and drinking treatment. The decrease in consumption is due to a decrease in the average length of stay at resorts to 15 days, and due to the appointment of pearl, coniferous, and artificially prepared mineral baths instead of using the abundant natural carbonic, hydrogen sulfide, and radon water of the CMW region in balneotherapy. At the same time, there is an increase in mineral water bottling. Of the currently extracted 3,000 m^3/day of all types of mineral water, only 1,500 m^3/day is used for drinking and balneotherapy at resorts, and the rest is used for industrial bottling.

5.2.8 *Novosibirsk Resort on Lake Karachi*

According to popular tradition, the soldiers of Khan Kuchum recovered their strength on Lake Karachi in the sixteenth century. As the hordes of nomad warriors moved quickly, talk of healing mud and water spread at the same speed. They gave the name to the lake. The lake was originally called Achu-Tebis, which is translated from the

Tatar as "bitter-salty," then it was renamed Karachi, which translates from the Tatar as "black silt."

The number of visitors to the lake increased from year to year. In 1880, two merchants, Norakovsky and Milinis, leased the lake with adjoining land for 12 years. Huts with plank beds, covered with turf, were built on the shore, where about 30 people could be treated at the same time. Then the number of beds increased to 65. There was no medical observation.

Among the locals, the lake was considered a deity. The village mullah, before allowing a patient to go into the mud, recited a prayer over him, in which he made assurances of the possibility of receiving relief.

A scientific study of the healing properties of the lake began in 1897. In the laboratory of the Tomsk Mining Administration, A.P. Bogachev made a chemical assay of the brine and mud. In 1901, L. Vertinson published the first information about the Karachi Lake resort in a guide to Russian resorts.

After the launch of the Siberian Railway, the railway department drew attention to the healing properties of the lake and built a bathing section and several wooden houses in 1900. In 1901, a railway resort was opened, where brine and mud treatment was carried out under medical supervision. The flow of patients increased every year. Entire streets of small private houses appeared on the lakeshore. In 1911 and 1913, the enterprising merchants Ketov and Scheglov opened two private resorts, and there were four private and one railway resort on the lake.

After the October Revolution, during the Civil War, the resort was badly damaged. All the buildings and half of the brine division were destroyed. In 1920, V.I. Lenin issued a decree on the nationalization of resorts. All resorts on Karachi Lake were transferred to Sibstrakhkassa. A few months later, the resort was operating at full force to receive workers and civil war veterans. In 1924, the restoration of the Lake Karachi resort began (Fig. 5.13).

In 1925, with the participation of doctors from the Tomsk Physiotherapeutic Institute, the resort was modernized. In 1926, under the guidance of Professor M.G. Kurlov, a lot of work was done to study the effect of therapeutic baths of Lake Karachi on the human body, as well as methods of treating diseases. From that moment, the Lake Karachi resort took first place among the balneological resorts of Siberia.

Fig. 5.13 Lake Karachi resort in the past

During the Great Patriotic War, a military hospital was created on the basis of the resort, which served as a rear hospital during the evacuation. Thousands of wounded recovered their health there. At this resort, doctor A.N. Bykhovskoy developed the original cold-shovel mud treatment method for gunshot osteomyelitis, which has been successfully used in many evacuation hospitals in Siberia and the Urals.

During the war, more than 800 tons (about 35 wagons) of therapeutic Karachi mud were sent to the country's hospitals. The methods of balneo-mud therapy, developed under the guidance of Professor I.A. Waledinsky at the Lake Karachi resort, made it possible to reduce treatment time for gunshot injuries of bones and long non-healing wounds from 170–200 to 70 days.

However, the resort lacked drinking water. In 1958, at a depth of 1,173 m, geologists discovered an aquifer with mineral water. Since 1960, Karachi water has been used in the resort as the third natural healing factor for the treatment of gastrointestinal diseases. A drinking gallery was built at the resort (Fig. 5.14).

Unfortunately, this unique resort was in disrepair for a long time, but in recent years it has been restored, and it has become a modern medical institution.

In 2013, a fourth healing factor appeared at the resort—mineral healing iodine-bromine water extracted from a depth of 2,660 m and used for bathing.

Fig. 5.14 Lake Karachi resort. Main building

5.2.9 *Resorts of the Transbaikal Region*

Darasun resort. The healing properties of springs in the village of Darasun in the Karym district of Transbaikal have been known since the end of the seventeenth century. The first medical facilities were opened in 1819; at the same time, the first residential buildings were built in the river valley. Wider use of the springs began in 1852. The first baths were built in 1857 for the daughter of the governor of Irkutsk.

Darasun has been a resort place since 1858. The village is located in the valley of a mountain river, at an altitude of 740–760 m, in the forest, 133 km from Chita (Fig. 5.15).

Kuka resort. The newest resort in Transbaikal, the Kuka resort, is located in the valley of the Kisliy Klyuch River, 69 km from Chita, at an altitude of 800 m. The forested floodplain river terrace of the Kisliy Klyuch River, stretching in a southeastern direction, goes around Kreselnaya Mountain. The open valley of the Bezymyanny stream adjoins from the south, and from the east and northeast, there is the wide valley of the Kisliy Klyuch River. The resort's surroundings are picturesque; it abounds in wild berries and mushrooms in summer.

The first description of the spring of the future resort dates to the middle of the nineteenth century. At the same time, barracks and a Buddhist temple were built by the locals near the spring. By 1919, 74 suburban-type houses had been built near the spring. The Kuka resort has operated year-round since 1934. During the war years,

Fig. 5.15 Darasun resort. Photo of 1925

a military hospital was deployed in it, and in 1947 a resort was organized for the treatment of patients with gastrointestinal diseases. Since 1992, the spring has been the property of ZAO Sanatoriy Kuka.

The main natural healing factor is carbonic ferruginous carbonic magnesium-calcium water. It has a high magnesium content, significant mineralization, and a very high concentration of carbon dioxide (to 4.93 g/l).

Yamarovka. Yamarovka spring is located in the southwest of the Transbaikal region, 360 km from Chita and 110 km northeast of the village of Krasniy Chickoy. The narrow valley of the Yamarovka River (the right tributary of the Chikoy River) is located at an altitude of 980 m, in the southern foothills of the Malkhan Range, protected from the wind.

The official start date for the spring's use can be considered 1854, when it was visited by a Kyakhta doctor Marcinkiewicz, who confirmed the high healing properties of its mineral water.

From 1971 to 1992, there was a trade union resort at the spring, where up to 300 people could vacation. Since 1999, the resort has been transformed into Yamarovka resort OJSC.

The main healing factor of the resort is the cold (4.4–5.0 °C), carbonic, weakly mineralized carbonic magnesium-calcium, slightly acidic water with unique healing properties.

Unfortunately, the resort is currently inactive, but its buildings and structures have been preserved.

Molokovka. The Molokovka resort is located in the valley of the river with the same name, 19 km south of Chita. It is located on the Chersky ridge (820 m above sea level), in a small hollow well protected from the winds.

The history of Molokovka water is linked with the merchant P.M. Mitrofanov, who owned a water plant located across the Ingoda River and operating on the basis of Molokovka mineral springs. After the establishment of Soviet power in Transbaikal in 1921, Molokovka was transferred to the Ministry of Health and used as a recreation center. At this time, the resort had up to 30 houses, which housed about 100 people. The resort functioned in the summer only. In 1927, the resort was transferred to the military department. Since 1941, the resort has been used as the rehabilitation department of the district military hospital, where fighters of the Red Army were treated.

There is a spring with excellent mineral radon, hydrogencarbonate, magnesium-calcium, carbonated water containing molybdenum, and arsenic, belonging to ultra-cold with temperatures 0.7 to 1.5 °C. In its chemistry, water is not inferior to Kislovodsk Narzan.

Karpovka mineral spring. The spring is located 25 km north of Chita at an altitude of 900 m in a pine forest.

The Karpovka spring has been used as the foundation of the physiotherapeutic hospital of Transbaikal Railway since June 1978.

The depth of the aquifer of carbonic water is 30–100 m. The water contains dissolved carbon dioxide in an amount sufficient for balneological treatment.

5.2.10 Saki Clinical Resort Named After N.I. Pirogov–Resort of Natural and Innovative Medicine

The Saki clinical resort named after N.I. Pirogov is an old health resort, founded in 1837, when, on the orders of N.N. Muravyov-Karsky, a branch of the Simferopol Military Hospital was opened to treat military personnel with healing mud from Lake Saki. During the Crimean War (1853–1856), directed by N.I. Pirogov, the healing properties of the mud and brine of Saki Lake were successfully used for the aftercare of Russian soldiers. It was the first world experience of mud and brine use at the stage of rehabilitation.

Today the resort is a multi-specialized diagnostic and treatment complex providing highly effective health resort treatment year-round to patients in 12 groups of diseases, as well as medical rehabilitation after somatic diseases, surgical interventions, and injuries, based on modern medical and information technologies. The main medical profiles are musculoskeletal, neurological, urological, and gynecological diseases. The resort receives more than 12,500 patients annually from 60 countries worldwide.

The basis of the resort's development strategy is the optimization of its activities and introduction of innovative medical technologies using the natural factors of the resort to increase the effectiveness and quality of treatment.

The health resort's treatment and medical rehabilitation programs use natural healing factors: the silt sulfide highly mineralized mud of Lake Saki, the activity of which is significantly higher than Dead Sea mud; Lake Saki brine, highly saline (I class) brine with a salt concentration up to 220 g/l; mineral thermal (45 °C) hydrogencarbonate-chloride-sodium water with low mineralization from its own spring; coastal-steppe climate with sunshine more than 2,500 h per year, average annual air temperature +11.8 °C, relative humidity 77%, and low rainfall.

The resort has a complex of pools: an open summer and an indoor winter pool; a pool with electric lifts for physically challenged people; and pools with mineral water from its own spring, intended for underwater vertical traction of the spine, therapeutic swimming, aqua aerobics, and hydrokinesitherapy (Fig. 5.16).

The resort has been certified for compliance with standards and has received numerous diplomas and awards. In 2015, it was among the 100 best enterprises and organizations of Russia and was a leader in the category of "best resort of natural and innovative medicine."

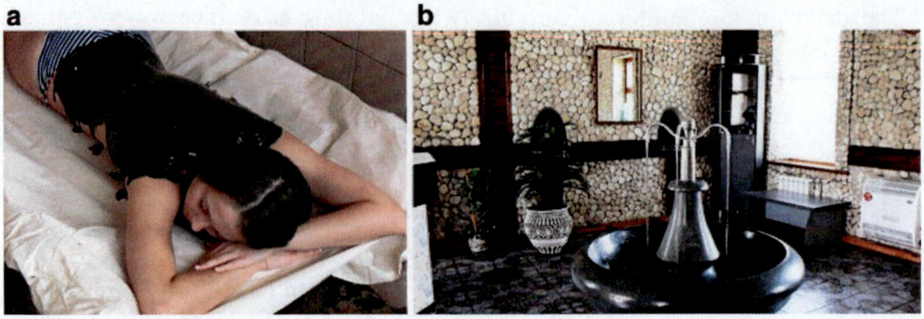

Fig. 5.16 Saki clinical resort–Mud treatment (**a**); Well room for thermal mineral water (**b**)

References

Federal law "On natural healing resources, medical rehabilitation localities and health spas" dated 23.02.1995 No 26-FZ. Legal reference system Consultant Plus (in Russian)

Kapusta LI (2006) Marcial waters. Pages of history of the first Russian resort. St. Petersburg: Dmitry Bulanin, 98p (in Russian)

Korovina VYa (2015) Walks in the resort of Staraya Russa. St. Petersburg: Aurora, 70p (in Russian)

Krivobokov NG (ed) (1994) Caucasian Mineral Springs. Moscow: Slovo, 304p (in Russian)

Narcissov AS, Narcissov DA (2018) Lipetsk: walks with the local historian. Lipetsk: Traditsiya, 72p (in Russian)

Schaefer GL (ed) (2017) Walking around the North Caucasus. Caucasian Mineral Springs. Travel Guide for pedestrians. Moscow: AST, 128p (in Russian)

Sheiko NI, Manshin NV (eds) (2011) Caucasian Mineral Springs. Moscow: Veche, 304p (in Russian)

Zhirova IA (2018) Lipetsk and its mineral waters. In: Lipetsk: walks with the local historian. Lipetsk, pp 51–61 (in Russian)

Conclusion

The territory of the Russian Federation is characterized by a multifarious environment that shapes a large number of healing springs and a variety of mud. The creation of the monography: *Healing Springs of Russia* is an important proactive step in the exploration of the country's natural healing resources, their use, and protection. The book will be an informational background for the examination of the potential capacity of ecological, medicinal, and spa tourism, health resort treatment, the development of the industry of water, mud, health-promoting, recreational, cosmetological, and other purposes. It will help to determine the state of exploration of potential natural healing capacity, identify regions where additional targeted investigations should be carried out as a priority, and facilitate the more effective use of natural healing resources.

The monography: *Healing Springs of Russia* is a scientific reference publication for a wide spectrum of users. The book reveals the country's healing abundance that could be used to improve public health. The information presented in the monography could also increase the tourism attractiveness of some Russian regions.

S. Malkhazova et al., *Healing Springs of Russia*,
https://doi.org/10.1007/978-3-030-83534-7

Natural bath in Pym-Va-Shor (Nenetz AO)

Sulfur lake (Samara Oblast)

Proval Lake (Pyatigorsk)